★ CITIZEN FARMERS

For Amiri
Thanks for your
great work! It's great to
have your rain barrels next door.

Farmer D
1/27/16

CITIZEN FARMERS

THE BIODYNAMIC WAY TO GROW HEALTHY FOOD, BUILD THRIVING COMMUNITIES, AND GIVE BACK TO THE EARTH

BY

DARON "FARMER D" JOFFE

WITH SUSAN PUCKETT

PHOTOGRAPHY BY RINNE ALLEN

STEWART, TABORI & CHANG   NEW YORK

Published in 2014 by Stewart, Tabori & Chang
An imprint of ABRAMS

Library of Congress control number: 2013945633
ISBN: 978-1-61769-101-0

Editor: Dervla Kelly
Designers: Heads of State with Danielle Young
Production Manager: Anet Sirna-Bruder

The text of this book was composed in Adobe Caslon, Chestnut, and Futura.
Text stock has been printed on 50% post consumer waste paper.

Printed and bound in U.S.A.
10 9 8 7 6 5 4 3 2 1

Stewart, Tabori & Chang books are available at special discounts when purchased in quantity for premiums and promotions as well as fundraising or educational use. Special editions can also be created to specification. For details, contact specialsales@abramsbooks.com or the address below.

THE ART OF BOOKS SINCE 1949

115 West 18th Street
New York, NY 10011
www.abramsbooks.com

★

CONTENTS

INTRODUCTION 6

1: COMPOSTING = STEWARDSHIP
20

2: PLANNING = VISION
50

3: TILLING = INITIATIVE
72

4: SOWING = FAITH
88

5: GROWING = PATIENCE
104

6: HEALING = COMPASSION
126

7: REAPING = GRATITUDE
150

8: SHARING = GENEROSITY
174

9: SUSTAINING = PERSEVERANCE
196

RESOURCES: FARMER D'S TOOL CHEST 218
ACKNOWLEDGMENTS 220
INDEX 222

"Urban conservationists may feel entitled to be unconcerned about food production because they are not farmers. But they can't be let off so easily, for they are all farming by proxy."

Wendell Berry

PARABLE OF THE CITIZEN FARMER

Here's a simple exercise to try at your office. Find a sunny window that people pass by often. Place a basil plant on the ledge or a nearby desk. Watch what happens.

A scenario could go something like this: A colleague wanders by the plant and pauses to inhale the intoxicating aroma. She asks if anyone's ever made pesto from scratch. Her desk-mate pipes up with his Italian grandmother's tried-and-true method from the Old Country. Later in the week, small jars of homemade pesto appear on everyone's desk. In time, that windowsill becomes crowded with other herb-filled pots. More stories are swapped, recipes shared, and gardening tips passed around.

Inspired, you try your hand at growing something edible at home. You start small, by planting a tomato and some herbs around your curbside mailbox. A dog-walker stops to chat as you're checking the mail, and instinctively you offer him a couple of beefsteaks and a handful of rosemary sprigs. The next day you find a basket of fresh eggs from his backyard henhouse on your front porch with a thank-you note.

The lady across the street observes the interaction and asks if you could use some of her excess zucchini. You propose a potluck dinner for all the gardeners on the block to share their abundance. At the party, a neighbor throws out the idea of turning the vacant piece of land at the end of the street into a community garden. You volunteer to seek permission from City Hall to use that public land, and before you know it, it's Saturday morning, and you and about fifty new friends of all ages are knee-deep in wood chips and compost.

Ideas and projects keep sprouting. You convince the PTA to start a schoolyard garden. You volunteer to help your church, mosque, or synagogue grow vegetables for its food pantry outreach program. You suggest planting herbs instead of annuals in those big planters on Main Street and maybe even starting an orchard in the park.

Through these volunteer efforts, your work skills develop, and you discover talents you never even knew you had. You feel happier and more invigorated than you have in years, and you find common ground with colleagues and neighbors you used to know only in passing. All because of that one little basil plant.

You do not need to trade in your urban clothes for overalls and move to the country to grow food. You don't even need to own green space. Across the nation, organic growers and consumers are bound by a common striving to leave the earth the way we found it—certainly no worse, and preferably even better. Take part in a community garden. Join a Community Supported Agriculture farm and volunteer to help with the harvest. Shop at your local farmers' market. Join a crop mob or another group of landless farmers who are lending their labor on weekends to farmers in need of helping hands.

This is what the "citizen farmer" movement is all about: taking actions that foster a healthier, more sustainable food system and passing on these values to the next generation. It is about honoring the place where you are now, believing in yourself and supporting others, sharing your wisdom and passion, and following your dreams. The steps outlined in the chapters that follow will help you create abundance in your garden as well as in your personal and professional life. Whether you support the movement from your garden, kitchen, classroom, boardroom, or farmers' market, I like to think of all of us as potential citizen farmers: each making a contribution to a better and more sustainable world.

I am convinced that integrating agriculture—and the personal virtues it teaches—into everyday life builds strong and vibrant communities. I believe this so strongly that I have made it my life's calling to cultivate as many citizen farmers as I can.

HOW I BECAME AN "ENTRE-MANURE"

You never know when you'll stumble upon something that will change your life forever. In my case, it was lunch: an innocent turkey and Swiss on rye.

I was a freshman at the University of Wisconsin–Madison and had started my day a lot like any other: I skipped class, rolled out of bed around noon, cranked up the reggae on the headset, and headed for this funky little café called the Radical Rye. As I ate, I found myself staring at my turkey sandwich: the tomatoes, the

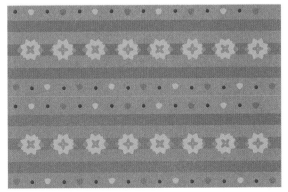

mushrooms, the lettuce, the onions, the turkey, the slice of Swiss cheese. Then it hit me: I had no idea how this sandwich came to be. Of all the meals I'd eaten in my lifetime, this was the first to stare back and cause me to ask it, "Where did you come from? Who raised you and how did they do it?"

Farming is not my heritage—far from it. I was born in the congested industrialized metropolis of Johannesburg, South Africa, on June 28, 1976. That same month a young boy named Hector Pieterson was shot when police opened fire on student protestors marching for equal rights, spurring the revolution that eventually led to the end of apartheid.

My grandparents and great-grandparents had immigrated to South Africa from Eastern Europe to escape the persecution of Jews under Stalin's rule. As the political situation in South Africa became more precarious, my grandparents encouraged their kids to make a new life in America. When I was a toddler, my parents took their advice and moved to Sandy Springs, a suburb of Atlanta, Georgia, where relatives had already joined a sizable South African Jewish community. My mom and dad inherited their parents' strong work ethic and love of family and community. They taught my sister Cindy and me to be humble and grateful, making sure we earned and appreciated whatever they gave us: a BB gun, a video game, a car, a college education, even a farm.

When it was time for college, I chose the University of Wisconsin–Madison, which I liked for its hip intellectual vibe and natural beauty. I signed up for the usual hodgepodge of liberal arts classes, but spent most of my time following the Grateful Dead and touring around national parks seeking community and inspiration. My dietary mainstays of chicken wings and microwave dinners gave way to tofu, tempeh, and sprouts. I hungered to do something positive for humankind, but I had no clue yet as to what that might be.

And then I took a hard look at that turkey sandwich. In an instant, the root of my dissatisfaction became clear: I was a consumer. I was taking more than I was giving back, and I had never really produced anything myself. Suddenly I had a burning desire to change that. I decided I wanted to learn how to grow a turkey sandwich from scratch. I wandered over to the agriculture school at UWM to see how I might pursue this rather obscure idea. I dug up an old file listing organic farms that offered internships and managed to convince a professor to sponsor me for a summer field study on one of those farms.

My first apprenticeship was at the Prairie Dock Farm in Watertown, Wisconsin, where I studied under a skilled farmer by the name of Greg David. Greg taught me the powerful lesson that if you have a vision and believe in it deeply, it will come true. By preparing the soil just right, and planting and nurturing seeds with good intention, I could bring the garden of my dreams to reality. His farm was a permaculture showcase of mobile chicken coops doubling as greenhouses, pigs turning compost piles, weed-seed-eating guinea hens, and abundant gardens and orchards.

Besides learning to drive a tractor and drag a hoe, I discovered that the ingredients for that turkey sandwich likely traveled more than three thousand miles, consuming their own share of fossil fuel calories and leaving a trail of carbon dioxide and pesticides behind.

Other shocking truths about modern industrial agriculture came to light. For every unit of food we consume, six times that amount of topsoil is lost. Small family farms are rapidly being swallowed up by industrial-scale monoculture farms. One of every three bites of food we take depends on pollination from a honeybee, yet pesticides, pollution, and the commercialization of the bee industry are causing the collapse of bee colonies worldwide. No more bees, no more tomatoes for that turkey sandwich.

In my search for answers about that fateful turkey sandwich, I found the career path I wanted to pursue. So on a trip back to Atlanta (in a caravan of VW buses going to see the Grateful Dead), I approached my parents with the idea of dropping out of school to continue my hands-on study of sustainable agriculture. After a passionate debate on the pros and cons of such a decision, my open-minded and supportive parents—somewhat skeptical but intrigued—decided to go along with this unconventional scheme. I remember my mother's reaction best: "Okay, Farmer D, let's see where this goes." And that's how I became Farmer D.

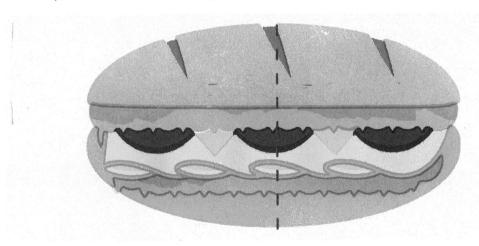

ORGANIC AND LOCAL TURKEY SANDWICH
- Low food miles
- Happy animals
- No harmful chemicals
- Dollars kept in local economy

INDUSTRIAL TURKEY SANDWICH
- Traveled thousands of miles
- Chemicals, pesticides, and antibiotics are unhealthy for people and animals
- Money goes out of region; supports middleman rather than farmers

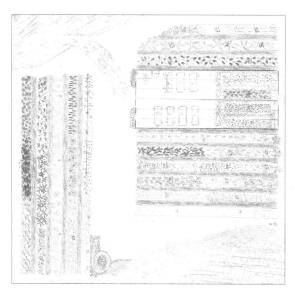

GROWING AGAINST THE GRAIN

I was determined to prove to my family and friends that I could follow my unorthodox passion and still be successful. In lieu of completing my college education, I apprenticed on biodynamic farms, attended sustainable agriculture conferences, read books, and talked to many farmers.

While I learned about the source of food, I also came to discover the soul of food. I had many teachers and mentors along the way. At a seminar on biodynamic farming, I became intrigued with the teachings of Rudolf Steiner, the early twentieth-century Austrian scientist and philosopher. Steiner advocated for a deeper understanding of the spiritual world and provided insights into how to incorporate a more holistic approach to education, agriculture, and many other areas. While he is best known as the founder of Waldorf education, biodynamic farming, and anthroposophical medicine, Steiner was also a leader in social reform, music, arts, and architecture. Steiner taught about ancient methods that nurtured the earth as a living organism, techniques that could be the antidote to the problems that plague modern agriculture all over the world. Bixodynamics, contrary to common misconception, is not all about farm gnomes and garden fairies. Most biodynamic farming lessons are no more mystical than the *Farmers' Almanac:* Plant and harvest with the cycles of the moon and stars. Raise animals, make compost, grow crops, replenish the soil. Produce rather than import inputs such as feed and fertilizer. These are the tenets of a well-managed biodynamic farm, which is essentially organic farming with an additional emphasis on sustainability and a spiritual approach to stewardship.

Miraculously, the only CSA (Community Supported Agriculture) farm in Georgia at the time was a biodynamic farm less than twenty minutes from my parents' cabin in the Blue Ridge Mountains. It was there that I met the farmer who would become my second mentor and true biodynamic guru, Hugh Lovel. I came to this farm, Union Agricultural Institute, to study for eight months. Every night, Hugh and I would discuss biodynamics over farm-grown feasts, and he would delve deeply into his philosophy of quantum agriculture. After this download, I would fall asleep to the gentle croaking of tree frogs and chirping of crickets. I was as happy as a proverbial pig in biodynamic manure.

The more I saw the fruits of my labor, the more convinced I became that food grown by these methods, without artificial chemicals and in tune with the rhythms of nature, was the future of agriculture. At the age of twenty, I bought—with my family's help—175 acres in southwest Wisconsin and started my own biodynamic CSA farm. In addition to honing my farming skills, I learned how to run a business, create partnerships, and manage a staff. The farm was such a success that in 1998 the American Biodynamic Association voted me "Rookie Biodynamic Farmer of the Year."

A BIODYNAMIC PRIMER

Many people over the years have asked me to explain what biodynamic farming is and how it is different from organic farming. This is not an easy question to answer, especially if they want an answer in just a few sentences.

Here is my very brief response if you have only 30 seconds:

Biodynamic agriculture is a type of organic farming that treats the farm as a self-contained living organism that can provide everything it needs from within. It is a closed-loop sustainable approach to agriculture that focuses on growing plants, feeding animals, making compost, and replenishing the soil. Repeating this cycle creates a regenerative process that improves the fertility of the farm over time rather than depleting it. Biodynamic farmers also apply homeopathic medicine to the earth by making special preparations to further enhance the quality of the soil, crops, and animals they produce.

Here is a more in-depth response for those with a little more time, best explained over a cup of tea and preferably with a notepad for some sketches:

Biodynamic farmers are constantly harnessing the forces of nature to help them grow vibrant, healthy food. All life basically depends on and grows toward the sun. There are macro and micro forces at work with the sun's influence that define the polarities at work on earth. Energy follows the sun; when the sun is on the other side of the earth from where we are, energy is being drawn downward—think of dew settling in the evening or leaves falling in the autumn, as the earth takes a big breath inward. This downward force is known in biodynamics as the earthly polarity; it is directly related to the water and earth elements and to lime and calcium. The opposing polarity, known as the cosmic force, relates to the fire and air elements and is seen in the upward working of silica. This force can be seen in the sunrise, rising dew, and bursts of growth in spring. It is the earth exhaling as energy streams up toward the sun.

As an example, horn manure, or BD (biodynamic) preparation 500, is used for enhancing the downward earthly working forces of lime, which stimulates digestion in the soil. It is made by stuffing cow (not bull) horns with fresh cow manure. The horns are buried in the fall and dug up in the spring. The preparation is then applied in the evening with the settling dew, especially during the fall when the soil is being tilled before planting a fall or winter crop. A small amount per acre is mixed with rainwater and stirred in a rotating vortex for one hour, which is a type of homeopathic potentization, and then sprayed out over the land. Another preparation, horn silica (BD prep 501), is used to enhance the upward, cosmic force of silica. It is made by stuffing cow horns with crushed and powdered quartz crystals and burying them from spring until fall. A small amount per acre is stirred for one hour and applied in the morning with the rising dew, especially in the spring.

These opposing but formative forces illustrate the earthly and cosmic influences at work in nature. Harnessed properly, these forces can be balanced and enhanced to aid biodynamic farmers in their goal of growing crops of optimum quality.

Biodynamic farmers care deeply about how their farming practices impact the environment and people's health. The philosophy and practices are intended to increase the quality and quantity of food grown while simultaneously enhancing the life in the soil and balancing the ecology of the farm.

THE JOHNNY APPLESEED APPROACH

Much as I loved rural living, I knew that isolation would not help me raise awareness among the masses. To spread the message about the benefits of local, organic, and biodynamic agriculture to more people, I decided to take the Johnny Appleseed approach.

After four years, I sold my farm in Wisconsin. I moved to California to run a farm in a youth prison. Then in 2001, I entered the Landscape Architecture and Ecology program at the University of Georgia–Athens. While there, I was hired to start and manage a CSA education and research farm called Full Moon Coop. In 2003, I received the prestigious Joshua Venture Fellowship for my nonprofit work as founder and director of Gan Chaim (Garden of Life). I approached the dean at UGA with the conundrum that I could not do it all. The dean suggested I withdraw from school and pursue these opportunities—and he even threw another one at me. He showed me the plan on his desk for a sustainable community just south of Atlanta called Serenbe Farms, which needed someone like me to help start their organic farm. I took his advice, and for the next three years I toiled the heavy Georgia clay soil and brought Serenbe Farms to life, while also starting farms and gardens at Jewish schools, camps, and community centers all over the country through Gan Chaim.

From then on, I began consulting for other projects and continued my quest to teach and inspire others to grow community through agriculture, while also expanding my entre-manurial endeavors. In 2006, I launched the Farmer D flagship product, the first certified biodynamic compost in the country, which was made with the spoils from Whole Foods Markets throughout the Southeast. Over the next few years, more products were added, including a planting mix, a fertilizer, and a line of cedar raised beds—all through a partnership with my father, a master woodworker and seasoned entrepreneur.

In 2008, many of my consulting clients were experiencing financial troubles and put their farming ventures on hold. During the next few years I developed an online and brick-and-mortar retail store selling organic gardening supplies: our signature Farmer D products as well as a wide variety of tools, seeds, plants, and much more. The store kept me busy building gardens and communities around Atlanta for several years, until my consulting business picked back up and I found myself flying all over the country setting up farms again. I have learned many things about gardening, business, and life through these incredible experiences, and I wanted to write this book to share the knowledge I've gained along the way. We can all be better stewards of the earth by growing healthy food, improving our lives, and creating a sustainable legacy for future generations. If each of us cultivates something and inspires others to do the same, then together we can grow into a worldwide coalition of citizen farmers.

FARMER D'S CITIZEN FARMER BASICS

1. MAKE COMPOSTING A WAY OF LIFE. Healthy soil = healthy food. Healthy food = healthy people. Healthy people = healthy communities.

More than 60 percent of typical "trash" that goes into landfills could actually be composted. I have been working with Whole Foods Markets in the southeastern United States to compost over four million pounds of food scraps a year into high-quality compost that is used to grow food at schools, homes, restaurants, and local organic farms. I also teach people how they can make compost themselves and offer many options that can be adapted to their personal situations.

2. START OR TAKE PART IN A GARDEN. Wherever you are, you can plant the seeds for growing a community by starting a garden at your home, in your school or workplace, or in your neighborhood. I have helped many cities grow communities through gardens. Among them was Harvest Farm, one of the largest community gardens in the country, in the city of Suwanee, about 40 miles from Atlanta.

For close to a decade, I have been working with private developments on incorporating organic farms and gardens into their master plans. One such project is the 10-acre farm at Natirar in New Jersey, the first American outpost of Sir Richard Branson's Virgin Limited Edition Collection of luxury retreats. The farm provides fresh produce, herbs, eggs, and meat for a culinary school, restaurant, and spa. I also helped design a 2,400-acre organic community in the Florida panhandle, anchored by a Farmer D farm. Throughout the country, communities are replacing golf courses with farms, cities are integrating urban farms into their downtown areas, and community gardens are popping up everywhere.

3. GIVE BACK. A big part of what it means to be a citizen farmer is sharing the harvest. Seeing the poverty and injustice in my South African homeland at an early age taught me gratitude for what we have and inspired

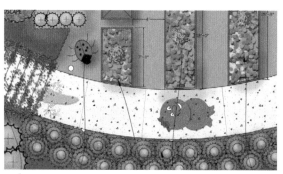

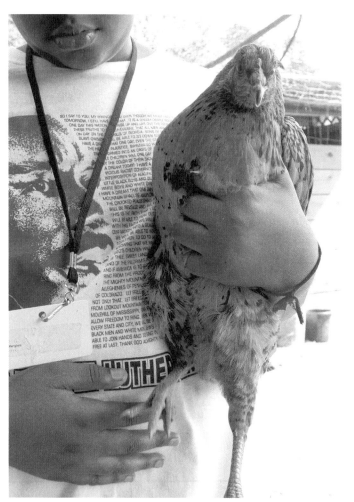

in me a desire to give back to those with less. Not everyone is lucky enough to have the land, resources, and ability to grow food. There are so many people around the world who go hungry every day, and so many others who are overeating food that does them more harm than good. We are frighteningly out of balance. To turn this around, we must think beyond our own dinner tables.

Of all my work, I am most passionate about the gardens we build for children and people in need. Over the years, I have built farms and gardens for a youth prison, a boys' home, a children's hospital, and various summer camps, including one for children with special needs. There are numerous ways to get involved, and in the pages ahead I'll show you how.

4. TEACH THE NEXT GENERATION TO APPRECIATE WHERE THEIR FOOD COMES FROM. There are so many lessons kids can learn in a garden—and often children will in turn teach their parents and help change family behaviors. My company has designed and planted more than one hundred gardens in schools. We believe every school should have a garden and we are working hard to make that happen. You realize how important these gardens are when you talk to kids who think carrots grow on trees. I can't tell you how many times parents and teachers have told me that kids feel differently about vegetables they grow, water, and harvest themselves—they actually want to eat them!

5. HAVE A SUSTAINABLE PLAN FOR THE FUTURE. No doubt about it: Making a living at farming is not easy. But it can be done. It is gratifying to see that, in recent years, growing numbers of twenty- and thirty-somethings—often college-educated, with no agricultural background—are choosing farming as a career. Like myself, these aspiring farmers have become disillusioned with industrial agriculture and are enamored with the prospect of becoming part of a broader movement to restore faith in the food supply and leave a positive legacy for generations to come.

READY, SET, GROW!

Each of the following nine chapters represents a step in the process of growing a healthy garden as well as a principle to live by. I hope you enjoy the personal stories I use to illustrate these values, tell you a bit about my journey, and share some lessons learned. I hope you will find the tips, tools, and techniques helpful for both your garden and your life.

I hope this book inspires you to be a steward, to plan to plant something, to get your hands dirty, to sow seeds, to grow food, to heal yourself and the earth, to reap the bounty, to share it with others, and to help foster a more sustainable future.

"Cultivators of the earth are the most valuable citizens. They are the most vigorous, the most independent, the most virtuous, and they are tied to their country and wedded to its liberty and interests by the most lasting bands."

Thomas Jefferson

1.

COMPOSTING = STEWARDSHIP

Every successful venture begins with a healthy foundation. Before we can grow food, we have to grow soil—and lots of it.

I make good dirt for a living. I also make it at home, every day, beginning in the morning when I toss the grounds from my coffee and the banana peel from my smoothie into the compost caddy by the sink. When it's full, I dump it into the backyard compost bin. Over a period of weeks and months, billions of bugs, bacteria, and fungi will convert these stinky, rotten kitchen scraps mixed with yard debris into fertile, sweet-smelling, life-giving earth.

Soil is the skin around our earth and the source of our sustenance, where life starts and finishes. Yet most of us take this vital resource for granted. Much of our earth has been degraded, overfarmed, and undernourished for many decades. We are currently losing topsoil faster than we can regenerate it. Millions of tons of organic materials that could be composted are taking up valuable space in landfills, rather than being returned to farms to grow healthy food.

Replenishing soil is just one of the many positive outcomes of composting. This practice can also drastically reduce, if not eliminate, the need for petroleum-based fertilizers, which pollute soil, air, water, animals, and humans. Compost builds the organic matter in soil, which helps plants grow strong and increase resistance to pests, disease, and drought.

Composting is a daily reminder of our individual tasks as citizen farmers: contributing to a healthy planet and being a responsible steward of the land. In this chapter, I hope to inspire you to make composting as essential to your daily routine as brushing your teeth. I want to share with you some very important tips on how to make your soil thrive so the food you grow helps you to thrive, too.

LOVE AT FIRST SCOOP

Until college, I never thought twice about tossing table scraps anywhere other than in a trash can or down a garbage disposal. It was shortly after my turkey sandwich revelation when someone I met at a party told me about an introductory workshop on biodynamic farming that was coming up at the Michael Fields Institute, about an hour from Madison. Founded in 1984, this nonprofit organization is one of the cornerstones of the biodynamic movement in the United States, with the mission of promoting sustainable agriculture through education, research, and outreach. There, I studied under several composting gurus who changed the way I thought about "dirt" forever.

Under their guidance, about a dozen farming newbies, including myself, learned how to build a biodynamic compost pile with garden refuse, dairy manure, spent hay, and a little soil from the garden. We spent a grueling day shoveling and spreading layers of this organic matter as if making a monster-size batch of lasagna. As the pile grew taller, we leaned a huge wooden board against the pile so we could run the wheelbarrow up to dump more materials on top. By the end of the day, our pile had grown to the size of a pickup truck. We covered it all with a layer of straw and doused it heavily with water, using a hose with a shower nozzle to replicate falling rain. Then we put holes in the pile at the locations specified in the biodynamic agriculture lectures by Rudolf Steiner, and added the requisite biodynamic compost preparations. The workshop was over, and now it was time to let Mother Nature work her magic.

I had dipped my toe into biodynamic farming and was now officially on my way to becoming a true citizen farmer.

DOWN THE RABBIT HOLE

Back at school, I scoured a list of local organic farms put out by the university's agriculture school and started calling them to see which ones offered apprenticeships.

Then I met Greg David, a former commercial builder turned organic farmer who occupied a bubble of biodiversity in the midst of an endless, sterile monoculture of corn crops about forty-five minutes from Madison. Greg hired me, and for the next six months I rose at the crack of dawn every day to drive from Madison to Greg's farm to work until dark—for fifty bucks a week.

Greg was especially passionate about restoring what had been a monoculture field of corn into a vibrant native prairie in order to create a sanctuary for wildlife in this agricultural desert. I helped him do this by preparing the ground and seeding it with a nurse crop of sunflowers that would provide shade and eclipse the weeds. Once the prairie was established, we would burn it periodically to maintain the revitalized soil.

Greg also had cutting-edge ideas about composting. He formed a relationship with the Department of Natural Resources and they would bring him truckload after truckload of fallen leaves and mucked-up lake weeds that would get dumped on the farm into long windrows. He was essentially harvesting all the organic matter from the region, composting it, and building up the soil fertility of his farm.

Once the mountains of leaves and lake weeds reached a certain size, he would fence his fifteen or twenty Tamworth pigs in around these piles. One of my jobs was to walk across each pile with a can of dried corn kernels and a shovel handle, and every 10 feet or so make a 3-foot-deep hole and drop a handful of corn down it. The pigs would root for the corn, and turn the compost in the process. When the pigs weren't being used to turn the piles, we would set them out to plow the fields by uprooting and eating the quack grass, which they could do better than any tractor. Not only did the pigs completely clean the fields of this pervasive weed, but they simultaneously fertilized the fields, saving us from driving the tractor all over them, compacting the soil, and burning fossil fuels.

I have a memory of walking the compost piles in the rain, collecting a bumper crop of the most beautiful volunteer tomatoes, peppers, and squash—plants that had sprouted voluntarily from the seeds of rotting fruit that had been turned by the pigs in this unbelievably fertile compost pile.

I had never before seen composting in action on that scale, and it was revelatory. This was biodynamic farming at its best, where animals were integrated into the agricultural organism in a humane and productive fashion, and the spoils of fallen leaves and rotting vegetables could be transformed into fertility for the land and an abundant harvest of food for the CSA.

BIODYNAMICS: A VEHICLE FOR ACTION

Armed with the basic tools for becoming an organic farmer, I was beginning to see a career path for myself that did not involve academia. After my awe-inspiring internship with Greg, I decided to withdraw from school and take on another apprenticeship instead, this time on a full-blown biodynamic farm in the North Georgia Mountains, run by Hugh Lovel.

Hugh had been farming this valley for more than twenty years and had turned a rocky creek bottom into one of the most fertile farms in the country—with compost, biodynamics, cover crops, and permanent grass and clover pathways between his vegetable beds. In the spirit of a true biodynamic farmer, Hugh produced everything he needed to sustain himself on his land, and I tried to follow in his footsteps. I managed the farm for most of my time there; I learned how to hand-milk cows and goats, make cheese and yogurt, grow rice, raise pigs, pickle okra, and save seeds. Hugh showed me how to build doughnut-shaped compost piles with manure and bedding from the animals. Compost from chickens would go to fruiting crops; cows, to leafy greens; goats, to herbs and flowers; pigs, to potatoes. I learned about radionics: influencing energetics through frequencies.

Prep making, planting with the moon, stirring, spraying, and sharing knowledge with others are some of the actions that biodynamic farmers take to make their farms and communities thrive. While many of these techniques may seem pretty unorthodox, one thing that is very clear is that the farmer's intention is the key to a well-managed farm. The goal is to be a good steward and leave the earth in better shape than we find it. Through Hugh, I became convinced that biodynamics provides a handful of very powerful tools for doing this and, at the same time, gives the citizen farmer's intention a vehicle for action.

COMPOST MEDITATION: COMPOST FOR THE SOUL

I have learned that, when applied metaphorically, composting can help us thrive in any environment we are planted in—be it the boardroom, classroom, or family dinner table. When building a compost pile, I like to turn the rhythm of shoveling into an internal meditation that allows me to compost negative thoughts into positive ones: fear into faith, judgment into compassion, insecurity into confidence, and anger into love.

Composting for the soil is the process of breaking down organic materials, while composting for the soul is the process of breaking down personal struggles (grudges, frustration, stress, anger). You bring the problems to the surface, add the good stuff to the heap (honesty, communication, support, nurturing), and give it time to "cook."

Building a compost pile is a meditation unto itself. Identify the location for the new pile by scratching a circle or square to mark the periphery. Loosen up the soil with a digging fork and add a sprinkling of lime, ash, and blood meal. As you build the pile, let your mind go and just tune into the rhythm of shoveling back and forth. It is an active meditation that results in a radiant pile of decomposing organic matter.

Even after this compost meditation is over, I use composting imagery frequently to dispel the mental clutter that may be holding me back in areas of my life beyond the garden. It has helped me make peace with business partners I've fallen out with, move on from stagnant relationships, remove obstacles impeding the creative process, and replace stress and anxiety with joy and gratitude—just like composting turns stinky raw manure and rotten vegetables into rich, sweet-smelling, fertile soil.

As the old Chinese proverb goes, a bad farmer grows weeds, a good farmer grows crops, and a great farmer grows soil. Be a great citizen farmer and make compost for a better life and a healthier planet!

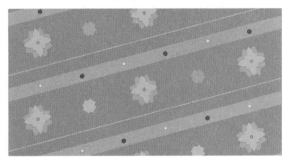

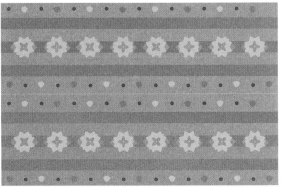

MOST COMMONLY USED TOOLS

Don't know the difference between a dibber and a digging spade? Thought a hori-hori was a type of sushi, not a garden tool? Never knew that a small hand shovel was called a trowel? Don't throw in the towel on tools! I'll help you learn what's what, and what you really need, from indoor gardening to market farming.

INDOOR GARDEN: If your idea of gardening is a few pots of herbs by the kitchen window, you really just need a dedicated spoon for adding a little organic fertilizer every now and then, and perhaps some little scissors for snipping herbs for your culinary creations or morning teas. (If you're into indoor hydroponics, however, that's a different story!)

BACKYARD GARDEN: If you've made the move outside your kitchen door, you could use a few gardening tool basics. These include a trowel, a garden fork (for tossing compost or digging potatoes), a spade, a hoe, a cultivator (that curved-tine tool), pruners, a wheelbarrow or big bucket for removing garden debris and moving compost, a garden rake for smoothing beds and raking leaves for mulch, and let's not forget gloves! In fact, two pairs would be best—one light pair for general garden work, and a heavier pair for handling things with thorns and tougher chores.

COMMUNITY GARDEN: If you're helping start a community garden, you'll need all the backyard tools listed above (and probably several of each) as well as basic building tools—a power drill, a saw, a hammer, a level, and a tape measure—to build garden beds, trellises, benches, sheds, and fences (add a post-hole digger for those). If you are growing food directly in the existing soil (rather than in raised beds), you will probably want to loosen up the soil with a tiller. Depending on how big your community garden is, you may find a direct seeder handy as well, especially if you are growing in long rows.

MARKET FARM: Once you move up to a market farm, where you are selling what you grow, you will probably have less community help and will want to automate things more. Now is the time for a tractor with attachments to create and turn rows quickly and easily, as well as specialized harvesting tools, depending on your crop mix and growing system.

Good-quality gardening tools and equipment are an investment. Get the best you can afford and take good care of them. Wooden handles need regular lubricating with linseed oil, mechanical parts need regular checking, and all tools and equipment should be stored properly so they don't age prematurely from exposure to the elements. See if your city offers free or affordable access to a shared tool bank, or work with your network to establish one so that more people can take advantage of helpful small-scale gardening and farming tools.

★

TOOLS FOR COMPOSTING

Composting requires a great deal of tossing and turning, as well as spreading and incorporating. Having a few heavy-duty tools will make this process less backbreaking and more productive and enjoyable. There are some very cheap, shoddy tools out there in the mass market that are tempting because of their price, but I encourage you to make an investment in quality tools that will last. You will find that not only is it a better investment in the long term, but it also makes the job at hand easier and more fun.

SHOVEL: For the most part, when making compost, any old shovel will do. For tumbler composting, I prefer a short-handled shovel to make it easier to get into tight areas, especially those with small openings. A long-handled shovel works best for bigger piles and wood bins. It is important to jab at the veggies in the compost pile to cut them into smaller pieces so they can break down faster. A standard hardware store shovel can cost between $15 and $40. If you want to invest in one that will last decades and inspire you to get more done in the garden, I recommend spending closer to $100 on a shovel with a forged-steel head and a sustainable hardwood handle with a T-grip.

COMPOST/MANURE FORK: Also called a hayfork, this tool is great for tossing and turning compost, as it allows you to grab a large amount of material per scoop. A digging fork, which is a more common home garden tool, will get the job done but will require more effort.

LAWN AND LEAF RAKE: A rake is useful for many tasks around the garden. A good leaf rake is a must for gathering the leaves that fall on your property so you can collect them for compost or mulch. It is also important to rake leaves so they do not kill off the grass below.

SOIL RAKE: A steel-head rake is useful for spreading compost out over your garden beds and for raking up mowed cover crops or garden refuse. This tool is great for preparing the soil in a bed before planting.

WHEELBARROW: There is no more efficient way to schlep compost and mulch all over the place than a good old-fashioned wheelbarrow. If a standard wheelbarrow is too much to handle, try a two-wheel wheelbarrow or a garden cart for more stability.

COMPOST CADDY: I know all of you are already collecting or are about to start collecting the kitchen scraps you accumulate from the yummy veggies you are eating every day. When you pick them yourself from your garden, there is even more to collect, such as the tops of carrots and excess leaves and roots that are cleaned off of grocery-store produce. An easy way to collect your kitchen scraps is on the counter near your cutting board and sink—this is where a compost caddy comes in. There are small, large, metal, ceramic, and even bamboo types available nowadays. I recommend metal or ceramic, definitely with a replaceable filter. If you eat a lot of veggies like we do, or have a family of four or more, I suggest a large size. These range from $20 to $60. For a more economical approach, a coffee can or bucket will do.

COMPOST AERATOR: Aerating your compost pile is important for speeding up the process and keeping the pile from going stagnant. While aerating can be done with a shovel or digging fork, a compost aerator is a handy tool for burrowing down deep in the pile with less effort. It is basically a long handle with some folding wings at the bottom. It is quite effective for home-scale composting and only runs about $20.

COMPOST THERMOMETER: If you are serious about making good compost, you need to get your pile up to at least 130°F. The best way to know if you are getting close is to use a compost thermometer, which is a thermometer on a long rod. A decent one costs between $20 and $30. You can also poke a deep hole into the pile with a tool handle and reach down with your hand to feel the heat.

WHERE TO COMPOST

Here are a few things to consider when planning where to put your compost pile.

SHADE: Compost is best done in a partially shaded area, especially during the hot summer months, to prevent the pile from drying out.

WATER: Place compost piles where they are exposed to rain and close to a water source, but not under a gutter, where they could get overwatered.

ACCESS: You will probably be visiting your compost pile on almost a daily basis, so put it somewhere close to the kitchen for easy access but don't get too close or you may attract fruit flies and other insects into your home. Also keep in mind where your finished compost will be hauled to—most likely your veggie garden, which is ideally close by and not up a big hill or stairs from your compost. Lastly, make sure there is enough room around your pile to get in with a pitchfork to turn the pile every month or two. If you use a tumbler, make sure there is plenty of room to load, turn, and unload.

ODOR: Sometimes compost can get stinky, so putting it near the front door or patio where you dine may not be a good idea. To minimize odor, add plenty of carbon and make sure the pile doesn't get too wet or dry.

CRITTERS: Where there is compost, there are likely to be all kinds of critters: flies, ants, roaches, rats, cats, and even raccoons. Keep your pile well balanced to minimize the odor, which attracts many of these visitors, and try to eliminate hiding places around the pile. Using a rodent-proof compost bin will help prevent many of these issues.

DRAINAGE: It is best to put your compost pile on well-drained soil.

COMPOST SYSTEMS

There is a composting solution for everyone. Nowadays there are many compost systems—small and large, simple and complex—for every situation. Here are some common composting systems to consider for your home, school, workplace, or community.

OPEN-PILE SYSTEM: The most basic option is not to have a bin at all, but instead use the carbon materials themselves to form a container. Hay bales work well, or you can just use straw or leaves layered with animal manure for the outside wall of the pile and put all your decomposing vegetables in the middle. Some people like to surround a pile like this with chicken wire or a plastic compost bin to keep it more contained.

WOOD AND/OR WIRE BIN: If you've got access to some old pallets, you can make a thrifty and functional

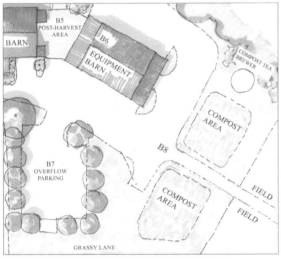

bin. In our wood shop, we build and sell bins that are both sturdy and stylish, using untreated FSC red cedar with a hinged lid and removable slats on the front for easy turning.

THREE-BIN SYSTEM: If you're impatient with how slowly the composting is going, consider trying a two- or three-section system. Toss your compost pile with a pitchfork from one open bin to the next every month or two in order to add air and to speed decomposition.

TUMBLER: If you want to step things up, consider getting a closed composter that spins. This is an easy, popular choice for home and school gardens, and an excellent way to compost kitchen scraps in an urban environment. These clever devices make turning the compost almost effortless. Rather than requiring a pitchfork to flip your pile from one bin to another, the tumblers are spun or rolled to mix and aerate the ingredients. I prefer the ones with two chambers; that way the material in one chamber, once full, can break down without new raw material being added. Another advantage to the tumblers is that most of them are rodent-proof. They should have ventilating air holes, which are important for good composting, but do not allow vermin to get inside.

WORM BINS: Worm bins are fun, easy, and productive, and the finished "vermicompost" does wonders for your garden (and boosts germination for certain crops like you wouldn't believe). Worms are little munchers, so they prefer their meals finely chopped. Juice pulp is caviar for worms! The problem some people have with worm bins is overfeeding the little guys. Yes, they eat half their weight in pounds per day, but it's best if you take it slow and see what that really means before laying out a daily buffet. There are many worm bin options out there—some homemade, and some worm condos that are pretty high-tech. We build wooden worm bins and also sell a lot of the stackable recycled plastic ones that have a spigot for capturing the worm tea, a nutrient-rich liquid that accumulates in the worm bin and is very good for your plants.

ELECTRIC COMPOSTERS: Marketed especially to urban dwellers, these plug in and provide automatic turning, heat, and aeration. They're pretty slick, and also pretty expensive (about $300). But if you live in a condo and have no outdoor space, they may be well worth the investment; they can go inside or on a patio. They can also be used for composting pet feces! Do not use that compost on your veggies, though—just on flowers and landscaping.

BOKASHI: This composting technique is gaining popularity. Bokashi is a culture that ferments food scraps into compost; it works wonderfully, especially in urban environments. It's as simple as a setting up a bucket or barrel with the bokashi culture and adding your food scraps.

GRUB COMPOSTING: This grub-based system uses black soldier fly larvae to convert kitchen scraps and manure both into usable compost and into feed for poultry, fish, and other animals. These larvae will often appear naturally in compost piles and can also be bred and harvested using systems designed for this purpose.

HUGELKULTUR: This age-old German technique utilizes woody debris to create a heap that is covered with soil and then planted. This is a great way to transform yard waste such as grass clippings, twigs, branches, and leaves into a vibrant raised garden bed. These beds hold moisture well and help with drainage, especially on compacted soils. They also serve as happy homes for beneficial microorganisms and thus healthy plants.

Some things you can throw into your hugel bed include:

- Branches of all sizes
- Leaves
- Grass clippings and hay
- Kitchen scraps
- Compost (even if not fully broken down
- Soil
- Bark or wood chips
- Newspaper and cardboard

Do *not* add any treated wood, plastic products, or any of the following types of wood:

- Black cherry
- Black walnut
- Black locust
- Cedar

To build a pile, simply layer the ingredients in whatever size and shape you desire, placing the bigger stuff on the bottom. Cover with soil and compost and pack the bed down as densely as you can—and you're ready to plant.

CURBSIDE COMPOSTING: If you happen to live in a progressive city like Portland, Seattle, or San Francisco, you can separate your organics, put them in a curbside waste bin alongside your garbage and recycling bins, and clean your hands of the composting process altogether. While this is a nice service, I still recommend making your own compost so you can complete the nutrient cycle in your own landscape and experience firsthand the exciting process that makes many a gardener proud.

LARGE-SCALE COMPOSTING: If you have a business such as a restaurant, or if you work in a school or hospital, there are plenty of composting options for you to get involved in as well. There are large in-vessel composting systems designed to be placed behind schools or grocery stores. Some cities have compost pickup services that take food scraps to a commercial compost facility, where they are composted and sold to farmers, gardeners, and landscapers. If you are building a pile in a large backyard or on a small farm, a rectangular pile 6 feet wide by 6 feet tall by 20 feet long is a good standard size and is perfect for one set of biodynamic compost preparations. If you want to build a larger pile, I recommend that each pile be 6 to 10 feet wide, 8 feet tall, and up to 200 feet long.

HUMANURE: For the hard-core composter who wants to let nothing go to waste, humanure is the result of composting human feces and urine with sawdust and paper or other carbon sources. So long as the compost process reaches temperatures that destroy harmful pathogens (above 130°F for at least 72 hours), this finished compost can be used to grow both edible and inedible plants. Well-composted humanure is not to be confused with sewage sludge, which can contain heavy metals and pathogens.

COMPOST RECIPE

All ingredients that go into a compost pile should be natural and organic; "if it grows it goes" is a simple rule of thumb to follow. The most available compost ingredients in an urban area are leaves and woody material (best shredded) for carbon, and grass clippings (without chemical sprays) and vegetable scraps for nitrogen. If you can get your hands on some chicken, cow, horse, or rabbit manure, that will help. Rabbits and chickens are great for backyards and both produce rich manure for composting. Rabbits are the easiest; you can put hutches above ground with worm bins below and create rich rabbit-worm compost in no time.

The ratio of carbon (C) to nitrogen (N) is very important for a healthy compost pile. The C:N ratio should be about 25 parts carbon to 1 part nitrogen. While mixing carbon and nitrogen materials for composting is a very basic concept, the science behind it is complex and dynamic. Carbon brings form to things in nature; think about your bones, or the structure of a plant. Nitrogen is more like the energy running through the forms: pulse, will, and desire. When rich carbon and nitrogen materials are combined with intention, the foundation for future life forms arises. But if they are left alone they can become toxic. For example, raw manure without a carbon source like hay can smell foul and pollute the environment. Good compost, however, does not smell bad. Likewise, when the human body is healthy it should smell only on the inside, not on the outside, as Rudolf Steiner explains in his Agriculture Course.

Not all ingredients for making compost have the ideal 25:1 ratio, so a little mix-and-match is necessary. If there is too much carbon, the pile will be stagnant, and if there's too much nitrogen it will get stinky and wet. Here is a simple way to balance the C:N ratio in your compost pile:

- High C:N ratios can be lowered by adding nitrogenous materials like grass clippings or manures.
- Low C:N ratios can be raised by adding carbon materials like dry leaves, wood chips, or paper.
- Piles in humid regions may need a little protection from the rain, such as being under a tree or canopy.
- Piles in arid regions, on the other hand, should be built with a concave top to catch as much water as possible.

WHAT MAKES COMPOST WORK?

While humans are the stewards who facilitate composting, it is the microorganisms—including bacteria, fungi, actinomycetes, protozoa, and rotifers—that deserve the credit for making the process happen. In order to create a healthy and productive workplace for our noble bugs, the following things need to be fostered in your compost pile:

- **CARBON**—energizes the pile and produces heat due to its microbial oxidation. Carbon materials are typically brown and dry.
- **NITROGEN**—feeds the microorganisms that are busy oxidizing the carbon. Nitrogen materials are typically wet and green; they also include colorful fruits and veggies and animal manures.
- **OXYGEN**—oxidizes the carbon and fuels the decomposition process.
- **WATER**—needed in the right amounts to maintain the microbial activity without causing the pile to go anaerobic.

COMPOST INGREDIENTS

Here is a general list of carbon, nitrogen, and neutral sources for your compost, as well as important guidelines on what should always, never, and sometimes be added to your compost pile.

ALWAYS COMPOST:

CARBON (BROWNS):

- **LEAVES** (most leaves compost easily, especially when shredded; avoid heavy, waxy leaves like magnolias)

- **BARK AND WOOD**, preferably shredded and/or aged

- **DRIED GRASS CLIPPINGS**, hay or straw

- **NEWSPAPER** (shredded; avoid slick color pages)

- **DRY WEEDS** (dry until brown first)

- **DRY CORN STALKS**, corn cobs, other dry stalks

- **CARDBOARD**, finely shredded

- **SAWDUST** and untreated wood shavings (in small amounts)

- **DRYER LINT** (moisten first)

NITROGEN (GREENS):

- **FRESH GRASS CLIPPINGS** (avoid grass sprayed with chemicals)

- **FRESH WEEDS** (avoid seeds, compost well)

- **COFFEE GROUNDS** and unbleached filters (worms love them)

- **GARDEN REFUSE** (dead plants, prunings, and the like)

- **MANURE** (chicken, rabbit, goat, sheep, pig, cow, horse, etc.)

- **SPENT GRAIN** from brewing beer

- **HAIR** (scatter to avoid clumps)

- **DISEASED PLANTS** (Use with care; if pile doesn't get hot enough it may not kill the organisms. If you're worried, play it safe and throw diseased plants away. If you do compost them, don't use resulting compost near the same plant family that was diseased.)

- **ALGAE**, seaweed, and lake moss

- **AQUARIUM WATER** (freshwater tanks only)

- **SOD** (Make sure pile is hot enough so grass doesn't keep growing, and be careful of the poly products and chemicals that are common in sod.)

- **URINE** (When you've got to go, save toilet water by feeding your compost; just don't startle the neighbors!)

NEUTRAL:

- **ASHES** from untreated wood (very small amounts only)

- **DISH WATER** (use natural dish soaps)

- **BEVERAGES** (with natural ingredients)

- **EGGSHELLS**, preferably crushed (calcium source, slow to decompose)

- **MILK**, cheese, and yogurt (place deep in pile to avoid attracting animals)

NEVER COMPOST:

- **PLASTIC** and inorganic materials, such as Styrofoam

- Anything with toxic chemicals, such as treated wood

- Anything containing heavy metals, such as wood with lead paint

- **ASHES** from coal

- **BIRD DROPPINGS** other than poultry (may contain disease organisms)

SOMETIMES COMPOST:

- **PINE NEEDLES:** They will break down, but very slowly, so add small quantities at a time and use the rest as mulch.

- **FISH SCRAPS:** May attract rodents and create a bad odor, but great if you can handle it.

- **MEAT:** Not recommended for the home composter, but fine for commercial composting.

- **FAT AND GREASE:** Not recommended for the home composter, but fine for commercial composting.

- **BONES:** Very slow to decompose, better in a fire and use the ashes.

- **HUMAN MANURE:** *See* Humanure (page 35).

- **ANIMAL REMAINS:** I know it is harsh, but animals do die, especially on a farm, and their remains make great compost so long as the piles are reaching the necessary temperatures (above 130°F) for extended periods of time.

- **LIME:** Though a little garden lime is a great compost activator, too much can have the opposite effect and kill the composting action. Add about 1 cup of lime to every 25 cubic feet of compost and mix well.

- **CAT LITTER:** If you use natural litter, it is okay to compost it for non-food-producing landscaping.

- **DOG DROPPINGS:** Okay to compost for non-food-producing landscaping.

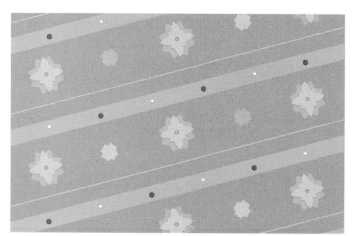

KNOW YOUR POO

In biodynamics, the farm or garden is treated as a self-contained living organism that provides its own fertility through animal manures and cover crops. Having the right balance of animals and crops helps ensure the soil is replenished year after year. Animals process grass into manure, which, when returned to the soil in the right way, provides it with the perfect medicine: nutrient-rich organic matter. Manure should be well composted before being applied to the garden.

This is a list of some common farm animals and what their composted manure is best for growing:

- **CHICKENS:** Chicken manure compost is high in nitrogen, phosphorus, and potassium, making it an excellent choice for fruiting vegetables like tomatoes, peppers, and squash.
- **GOATS:** I have found goat manure to be most effective on herbs and flowers.
- **PIGS:** Pig manure is perfect for potatoes and leeks due to its high potash and phosphorus contents.
- **COWS:** Cow manure is great for growing just about everything due to its balanced nutrient content, but in my experience it is best used for leafy green vegetables such as lettuce, kale, and collards.
- **RABBITS:** Rabbit manure is high in nitrogen and is also good for leafy greens.

COMPOST BOOSTERS

If you find your compost is not hot to the touch or a thermometer shows a temperature below 110°F, you should either turn the pile or add a nitrogen source or other biological stimulants to get it kicking. You can use any high-nitrogen organic fertilizer like blood meal, feather meal, or chicken manure to help give your compost a boost. You can also find some good prepackaged options—labeled as either "compost starters" or "compost activators"—at a local garden center, in a hardware store, or online. Another option that has been growing in popularity is bokashi, which is a fermented compost activator made from inoculating wheat germ or sawdust with Effective Microorganism (EM). When added to compost, bokashi helps speed up the decomposition process. My favorite boosters are the biodynamic Pfeiffer Compost Starter, backyard chicken manure, and homegrown comfrey.

BUILDING THE PILE

Start by identifying the area where you want to build your pile. With a digging fork or rake, scratch out a circle, square, or rectangle to mark the periphery of the new pile. Loosen the soil in that area and begin by layering browns and sunflower stalks, thistles, and other plants with hollow stems, but nothing woody.

Start building up the outer periphery with a 1-foot-wide band of material until it's about 1 foot high. I like to build piles in circles and usually go with a 5- to 7-foot diameter. Once you have built a base that is 12 to 20 inches wide by 1 to 2 feet tall, you are ready to start filling in the center a little bit at a time, making sure the outer wall is always at least 12 inches taller than the doughnut hole in the middle. This allows you to build a tall, strong pile. Here is a general sequence of materials to be layered:

- **CARBON / BROWN MATTER**—12 inches
- **NITROGEN / GREEN MATTER AND MANURES**—6 inches
- **SOIL**—2 inches

Add a layer of food scraps as part of the green matter, but keep these to the middle of the pile to avoid attracting animals.

As you layer the ingredients, use the back of your pitchfork or shovel to tamp the sides and top of the pile. Compressing the pile so the ingredients are dense helps get the pile cooking. You can hop on top of the pile and do a little "tamp dance" and then gently but thoroughly shower the pile with water.

ADDING THE BIODYNAMIC COMPOST PREPARATIONS TO THE PILE

Just as people add vitamins to their diets, biodynamic farmers add herbal preparations to their compost. Herbs such as chamomile, dandelion, yarrow, stinging nettle, and valerian are prepared according to the recipes in Steiner's Agriculture Course to best impart their healing properties into the compost pile. When these potent herbs are added to the heap, a certain physical and energetic healing takes place, enlivening the compost with microbes and forces that are beneficial to soil, plants, and animals. You can make biodynamic compost preparations at home or on your farm (see page 219 or refer to a local, national, or international biodynamic association) or a set can be purchased for about $25 from a few sources including local biodynamic farmers and the Josephine Porter Institute. One set of preparations (or "preps") can treat up to fifteen tons of compost, which is plenty for a very large backyard pile.

When your compost heap is built, follow these simple steps:

- Make five holes in the top of the compost heap, about halfway into the depth of the pile, distributed in an "x" pattern like the shapes on a number five playing card. The instructions that come with the preparations will provide more detail on how and where to insert the preps into the pile.

- Put 1 gram each of the BD 502-506 preps in the holes about halfway down into the pile, using one prep per hole. Fill in holes.
- Make a hole at each end of the heap. Stir 10 ml of BD 507 (valerian) in 1 liter of water for 10 minutes and pour half of the liquid into this hole.
- Sprinkle the remaining liquid evenly around the top of the pile in a clockwise direction.
- Spray pile with BD 508 (fermented equisetum tea) and toss equisetum plants on top of the pile.

Maintain the pile:

- Measure the heat periodically: Insert a compost thermometer or dig into the heap with your hands or a shovel to see if it's steaming.
- Water regularly.

TENDING YOUR COMPOST PILE

I'll tell you a simple secret: Compost happens. Things in nature decompose. If you pile some leaves and plant cuttings in the corner of your yard, they will break down naturally and you will have compost one day, whether you toss it every week or leave it alone. The difference will be the speed with which this happens and the quality of the finished product. Keep your compost pile moist and aerated by watering it when it is dry and turning it when it becomes wet or stagnant. If your pile gets hot, it will break down faster and will be less likely to have weed seeds or disease. If it stays cool, introduce worms and give it some more time to decompose.

TROUBLESHOOTING

1. COMPOST PILE NOT HEATING UP: To achieve the ideal temperature for quick composting, and also to kill weed seeds, you need at least a 3-foot-by-3-foot pile, with the right ratio of carbon (about 70 percent) to nitrogen (about 30 percent), plus sufficient air (turn the pile once a month) and water (slightly damp, not soaked). If your pile is not producing at the rate you want, you may need to adjust these components to get your desired results.

2. COMPOST PILE NOT BREAKING DOWN: All garden debris will eventually decompose, but if the pieces you include in the pile are too big, it might take a very long time. You can help the process along by chopping up veggies, big stalks, or branches and shredding leaves before adding them. Many people actually have a separate pile, such as a brush pile or hugelkultur bed, for larger debris like tree branches.

3. ODORS AND RODENTS: Tossing kitchen scraps in an open compost pile without burying them is an open invitation for every rat, squirrel, and raccoon in the vicinity to come for a feast. Make sure to cover food scraps with a thick layer of carbon material each time you add to your pile. But even if you do bury food scraps, some pests may still find their way to your pile. Investing in a closed, rodent-proof composter for your veggie and fruit peels is one solution. And, again, balancing the greens and browns will also reduce odor.

4. MAGGOTS: Disgusting though they might be, maggots—the larvae of flies—are generally considered harmless and may actually be beneficial (especially the larvae of the black soldier fly) in helping reduce the volume of a manure pile. Some species, however, can be carriers of disease. Maggots thrive in excessive moisture, and if you're using a closed bin without sun exposure, it's very difficult to dry it out. Mixing in dry high-carbon material such as shredded cardboard helps, as does adding bone meal.

5. ANTS: Ants in the compost pile could be a sign that your compost is too dry. Sprinkling cinnamon or turmeric is one deterrent. Add water to the pile and make sure your ratio of browns and greens is balanced.

6. WORM BIN GONE BAD: If your worm bin smells, it probably means you're overfeeding the worms and need to cut back on adding scraps. Another issue is worms that try to get out. This may be due to insufficient air (there should be airholes) or excessive moisture, which will basically drown them. It's normal to give your worm bin an occasional reboot with a deep clean and fresh bedding.

WHEN IS COMPOST READY TO USE?

Healthy, fertile soil should feel soft, light, moist but not wet, and crumbly in texture. It should smell earthy and sweet, with minimal to no ammonia odor. It should be teeming with life—in the form of worms, bugs, and microorganisms. If it was made with food scraps, the only things you should be able to distinguish are bits of eggshell, sticker labels, and avocado or mango seeds.

BUYING COMPOST

When buying compost, remember that not all compost is created equal. Be careful of poorly made compost, which can be riddled with weed seeds, pathogens, diseases, heavy metals, and even human manure. It's important that you know your composters and what their practices are to ensure that they are making a clean, high-quality product. Read the ingredients list and check to see if it is OMRI (Organic Materials Review Institute) certified or approved for use in organic agriculture. This means the compost has to have reached temperatures of

over 135°F for a certain period of time, and that the finished product has been tested for pesticides, heavy metals, and other possible contaminants.

HOW TO USE COMPOST

Whenever I grow a garden or farm, there are a few ingredients that are essential to getting started. Once the chemistry of the soil is balanced using lime and other minerals, I add compost to feed the soil biology. This enhances nutrients, moisture retention, disease resistance, and overall plant vitality. If your compost is not fully ready but you need the bin space to start a new batch, empty the contents into a container with drainage and let it cure outside for a few weeks until ready for the garden. The best times to apply compost in the garden are late fall and early spring, as well as in between crop cycles.

Here are the primary ways I use compost in the garden and on the farm:

· Add finished compost to garden beds in between planting cycles and incorporate into the top 4 to 8 inches of the bed.

· Use finished compost as side dressing around the bases of plants, then water and mulch. This can be done for landscape plants and lawns as well as veggies, herbs, and fruit trees. The bigger the plant, the more compost should be used.

· Use finished compost to inoculate a new or stagnant compost pile. Add a few shovelfuls to help kickstart a new pile.

· Make compost tea and apply as a soil drench or foliar spray. Opposite is a compost tea recipe that can be made in small batches for home use as well as in a large-scale compost tea brewer for commercial applications. There are a few caveats before you begin: When making tea, be sure to use good compost and follow the National Organic Program's safety guidelines. While unlikely, there is a chance that ingesting crops that have been treated with compost tea made incorrectly could lead to illness caused by coliform bacteria or salmonella.

GROW YOUR OWN FERTILIZER:
COVER CROPS AND GREEN MANURING

Between crop cycles I plant cover crops to let the land rest and to replenish nutrients and organic matter that were extracted by the previous crop. Once the cover crop reaches its highest nitrogen-fixing stage of growth,

★

FARMER D COMPOST TEA

MAKES 5 GALLONS, ENOUGH FOR A LARGE HOME GARDEN

1 CLEAN 5-GALLON (20-L) BUCKET (PREFERABLY CERAMIC OR STAINLESS STEEL. PLASTIC IS OKAY. NO BLEACH OR CHEMICAL RESIDUES)

4 GALLONS (15 L) RAINWATER, SPRING WATER, OR TAP WATER

1 QUART (400 G) AGED COMPOST

PANTYHOSE, CHEESECLOTH BAG, OR FINE-SCREEN COLANDER

STINGING NETTLE OR BURDOCK LEAVES

3 TABLESPOONS COMPOST TEA ACTIVATOR (SUCH AS UNSUL-FURED ORGANIC MOLASSES OR A READY-MADE ACTIVATOR SUCH AS GROWING SOLUTIONS COMPOST TEA CATALYST)

SMALL FISH-TANK AERATOR PUMP WITH TUBING AND SEV-ERAL BUBBLERS OR AIR STONES TO PUMP OXYGEN INTO THE BUCKET OF WATER WITH COM-POST AND ACTIVATOR

Fill the bucket with water. If using tap water, let it sit out for 24 hours to release chlorine. Place the compost in the pantyhose, cheesecloth bag, or colander. Fully immerse the compost in the water without allowing soil particles to get in the water. (You can also brew without a mesh bag and instead run the finished tea through a fine-mesh or cheesecloth strainer.) Toss in a few handfuls of fresh stinging nettles, burdock leaves, or other apropriate herbs and weeds for added benefits. Add the compost activator, turn on the aerator pump, and let brew for 24 to 48 hours. If you are not using an aerator, let sit for 5 to 10 days to allow for adequate fermentation.

Once brewed, compost tea has a short shelf life, so it should be applied as soon as possible, no later than 24 hours after being brewed. When using compost tea, dilute at 10 parts water to 1 part compost tea and apply as follows:

FOLIAR APPLICATION: Spray in morning or evening on underside of leaves on plants, shrubs, and trees.

SOIL DRENCH: Water new transplants or existing plants.

LAWN: Spray or run through your irrigation system.

around peak flowering time, I will usually turn it back into the soil, which is called "green manuring." Organic matter recycles into the soil and feeds the compost microorganisms and worms, which in turn feed the next crops planted here. This amazing process, like composting, is a natural, easy way to regenerate the soil and ensure a healthy garden for posterity. If you turn under the cover crop in the bed, wait two weeks to plant so that the organic matter has time to break down. Instead of turning under the cover crop, you can also harvest it and put it into your compost bin so you can plant in the garden right away.

Some of my favorite cool-season cover crops are rye, crimson clover, winter peas, and hairy vetch. For warm-season cover crops, I like iron and clay peas, Sudan grass, buckwheat, soybeans, and sunhemp.

COMMUNITY COMPOSTING

Composting is a great way to practice your citizen farmer skills by inspiring your neighbors and local business-people to follow your example.

Take note of those bulging leaf bags along your street. Next time you see the neighbors who put them there, say hello and ask if they'd mind you taking it off their hands for your home compost. You may find yourself sharing tips on composting, organic lawn care, and vegetable gardening. A nice way to say thank you—and maybe recruit some new citizen farmers—would be to bring them back a small sample of finished compost to use in their gardens or planters.

You can raise awareness simply by initiating the conversation. Ask your friends, local chefs, and coffee-shop owners if they are composting their food scraps and coffee grounds. If they are throwing away their organic material, offer to take it off their hands (maybe even for a small fee). Give them a special bucket and have them fill it with organic material only (I recommend taking them the list of compostable materials from pages 38–39, or—better yet—giving them a copy of this book!). Every few days have them leave the bucket outside where you can easily pick it up. You can either take the bucket and leave them a clean one, or dump the contents of the bucket into a large trash bin in your vehicle, wash out the bucket right there on site, and leave it for them to fill up again. Bring the food residuals back to your home compost pile and add lots of carbon, such as shredded bark, sawdust, or all those leaves you've collected from the neighbors. This could turn into a little business before you know it!

Work together with neighbors and friends to collectively compost at home, at school, at work, or in a community garden. Building, turning, and spreading compost is best done with other citizen farmers!

★ TOP 10 ★

PLANTS FOR MAKING COMPOST

Some crops have more left over for the microbes than others. These are some of my favorites for feeding the compost pile:

1. **COMFREY:** If there is one crop you should grow specifically to make better compost, comfrey is it. It is a hardy perennial and medicinal herb that enriches and speeds up decomposition when added to a compost pile. It can also be used as mulch beneath your veggie plants and as a slug trap—place comfrey leaves around the garden near slug-prone crops.

2. **PEAS:** While it is great to grow peas for their delectable little pods, their other asset is the plant and nitrogen-fixing roots that remain after you've harvested all the peas. Cut the plants above the root and toss it into your compost pile.

3. **CARROTS:** Most people grow carrots for their sweet, nutritious roots, leaving the large, leafy tops as perfect fuel for the compost pile. This is an excellent use for those vitamin- and mineral-rich greens, unless you feel adventurous and want to chop them up finely and add them to a salad or coleslaw. Some other options are to make a carrot-top pesto, whip up a soup, or add the greens to a batch of gumbo z'herbes.

4. **TOMATOES:** At the end of each summer season, tomatoes are likely to be the tallest, leafiest mess left in the garden. The semi-woody stems make a great base for a compost pile so long as they are not diseased. (It's better to play it safe with diseased plants and throw them away.)

5. **SWEET POTATOES:** All the vines that are left after harvesting a crop of sweet potatoes are an excellent addition to compost, as the greens are high in vitamins and minerals. You can eat the tips of the greens by picking the top 3 to 5 inches of the vines and adding them to a stir-fry or soup. If the stems don't snap off easily, try moving closer to the tip until they do.

6. **BEANS:** Both bush and pole beans have a good deal of plant matter left after harvest. As with peas, it is best to leave the nitrogen-fixing roots in the soil and add the rest of the plants to the compost pile.

7. **STINGING NETTLE:** While you may not want to plant this everywhere in your garden because of its stinging spines, you will definitely benefit from having a patch you can pick for food, tea, and compost. Stinging nettle is extremely high in iron, magnesium, calcium, and vitamins, which are imparted to the soil through compost made with nettles. Also makes a fantastic compost tea that can attack a pest outbreak if applied at first sighting.

8. **HAIRY VETCH:** A great cool-season cover crop that fixes nitrogen in the soil and produces a great deal of organic matter. I usually grow hairy vetch together with rye and winter peas, which together make a rocking green manure or nitrogen layer in your compost pile.

9. **GRASSES:** Growing perennial grasses in your pathways and lawn and annual grasses in your garden beds is a good way to draw nutrients up from deep in the soil, prevent erosion, and have a nice continuous harvest of grass and clover clippings to feed to animals and/or the compost pile.

10. **SUNFLOWERS:** Sunflowers in the garden provide both beauty and function. While they are very useful for trellising crops like cucumbers and pole beans, they also grow tall and fast, providing lots of organic matter for compost. Their big stalks are best used at the base of a compost pile; crosshatch them to help draw air up into the heap.

"A vision without a plan is just a dream. A plan without a vision is just drudgery.
But a vision with a plan can change the world."

Old proverb

2.

PLANNING = VISION

Over the years, I have been involved in helping plan farms and gardens in everything from backyards to master-planned communities. With each experience, I have sharpened my vision by trying to learn from my mistakes as well as my successes. If you are reading this book, you likely have a vision for a better world and want to do something to help make that vision a reality.

I suggest coming up with a plan of action that defines where you want to end up—and, more importantly, where to start. You've decided to plant a garden—but what do you want to grow? What are the conditions you have to work with and how much time and resources are you willing to invest? This chapter is an exercise in creativity and long-range "visioning," helping you answer these questions so that you can set clear and manageable steps by which to reach your gardening goals. I will walk you through my process for designing various types and sizes of gardens. I will offer tips for getting organized, drawing plans, making plant choices, and sustaining your enterprise.

BUYING AND PLANNING MY FIRST FARM

Early on in my biodynamic studies, I fell in love with the idea of having a farm of my own, where I could invest in the long-term fertility of my own land and plant fruit trees knowing that I would be around long enough to reap the benefits. That was right around the time I met James. He was one of a bunch of dreadlocked hippies I was living with in a funky bungalow near the campus of the University of Wisconsin–Madison. Even though he was about ten years and a hundred Grateful Dead concerts ahead of me, we quickly became very close friends. We shared a passion for farming, and together we developed a vision for how we could make this dream of becoming successful organic farmers a reality.

On weekends we would pack picnic lunches and drive down country roads, pipe-dreaming about starting a CSA farm. Our goal was to get together the following year and try to find a piece of land where we could set down roots and start a farm. I went off to work on farms in Santa Cruz, California, and then to apprentice for eight months under Hugh Lovel in the North Georgia Mountains; James stayed in Madison to finish his horticulture and landscape architecture degree.

As planned, James and I reunited the next spring and picked up our search for the perfect farm. We fell in love with the Kickapoo Valley in the southwest corner of Wisconsin, which is known as the Driftless Region. We planned budgets, sourced equipment, bought tools, and ordered seeds, knowing we would at least rent if we couldn't find something to buy.

We eventually fell upon a farm for sale—a diamond in the rough hidden under garbage, old cars, and 20-foot-tall ragweed, with about ten menacing dogs chained to trees in the yard. The fact that it had been neglected for decades made it a prime candidate for instant organic certification. We dug some soil samples beneath the ragweed and were amazed to find 2 feet of black, rich topsoil. This land was the most fertile we had seen, and with 2,000 feet of trout stream meandering through the valley, 20 acres of bottomland, 100 acres of forest with mature sugar maple, a 50-acre ridgetop field, and a cluster of useful buildings, it more than fit the bill. We yearned to clean it up and bring it to life. The only thing in our way was a mean landlord and a bank loan for $180,000 that we had no idea how we were going to get.

James and I decided to make an offer anyway. I went to one bank after another trying to get a loan. Most rural banks were used to funding conventional monoculture farms, so when I proposed 150 different crops, their eyes rolled back. Not wanting to miss another season, I negotiated to rent the farm with an option to buy, contingent on getting financing. After a dozen denials, I finally found a banker willing to give me a loan for 80 percent of what we needed, and somehow convinced the owner to give us a second mortgage for the other 20 percent. I sold the Coca-Cola shares my uncles had given me as a bar mitzvah gift, which served as the down payment, and my generous father guaranteed the loan from the bank.

Buying the farm was the most exciting, terrifying, and rewarding feat I had ever accomplished. By believing in something with all my heart and cultivating support from family and friends, I made a seemingly impossible vision come to reality. We had bought the farm, and now the real work could begin.

★

PLANNING MEDITATION

Before I start projecting my ideas for a new farm or garden onto paper and ultimately on the land, I always like to take a walk through the field where I will be planting and gently feel the spirit of the place. You'd be amazed at how this relaxing and nearly effortless practice can open your heart and mind to possibilities you may not have otherwise foreseen.

To do this, simply let yourself wander, guided by your intuition, with no agenda other than to experience where you are with keen awareness. Walk slowly and breathe gently. Let go of all your worries, projections, and negative thoughts. This, in my experience, is the space from which the best ideas come, and can be applied to any decision you make.

Rudolf Steiner suggests a meditation specifically for farmers in his Agriculture Course:

> **WHEN A PLACE CALLS TO YOU FOR A REST, SIT OR STAND AND LET YOURSELF FEEL INTO THE SOIL, THE GRASSES, THE TREES, THE BIRDS. VIEW THE SPACE FROM MANY DIFFERENT PERSPECTIVES AND LOOK CLOSELY AT THE DETAILS SUCH AS THE SOIL AND INSECTS, THE QUALITY OF THE PLANTS AND WEEDS GROWING THERE. BECOME ONE WITH THE LAND, THE PLANTS AND THE AIR. WITH LESS AND LESS EFFORT YOU ARE ABLE TO FEEL VERY LITTLE SEPARATION BETWEEN YOU AND YOUR SURROUNDINGS.**

This is what it's like to experience being a human "being" rather than being a human "doing." Learning to listen to the land—and allowing it to speak through you—is fundamental to successful biodynamic farming and a habit that can give you clarity in whatever endeavor you wish to pursue. Once your mind is open and content, you can set your vision on the garden of your dreams.

STEPS FOR PLANNING A FARM OR GARDEN

1. DO YOUR HOMEWORK. Read books. Volunteer at a local organic farm, take part in a crop mob, or join a CSA as a working member for a season. Attend conferences, workshops, and seminars.

Most regions of the country now have major organic farming conferences that the public is invited to attend. If you can't make it in person, there are recordings of past workshops and online webinars.

When you go to a farmers' market, ask the farmers questions, as they are usually more than happy to share their wisdom. If you want to get more serious, you can enroll in some classes or try a seasonal apprenticeship on an organic farm near you. If you like to travel, there is also an international internship program called Willing Workers on Organic Farms, through which you can work for a short spell on an organic farm somewhere exotic. Hire a consultant to help you avoid mistakes and make the most of your investment.

2. CULTIVATE A VISION. If you are working alone, start thinking about how much you can take on by yourself. Gardening is more fun as a social activity, so see whom you can get on board with your vision and get their input, too. This could be your spouse, your kids, a roommate, neighbors, a friend, or a local organization.

3. DRAFT A PLAN. Lay out a plan for what you are going to do when, and what you are going to plant where. Growing food requires good timing and attention to detail. The more organized you are, the easier it is to get things done efficiently.

4. WRITE UP A BUDGET. It is helpful to make a simple budget for what your garden is going to cost. You will need to invest in things like soil amendments, seeds, plants, and more. If you are interested in quantifying how economical your garden is, track how much you spend and how many pounds of produce you harvest (see pages 162–163 for ten crops that cut grocery bills). This will help you determine your return on investment for different crops and figure out how to get the most value from your garden.

5. RAISE FUNDS. Plan ahead by investing some of the money you will inevitably spend at the grocery store and use it for your garden instead. You can look at it as if you are prepurchasing your groceries, sort of like a CSA. If you are gardening on a larger scale than for yourself alone, you may even want to consider a small neighborhood CSA where each family pitches in for a share in the harvest. They may also have the option to pay less money and put in some time in the garden.

6. LOOK FOR LAND. First look around your home and see how much land you can grow on; if you need more than that, there are other options. You may be eyeing a neighbor's south-facing front yard as you walk the dog, for example, or an open green space at a nearby school or park. Talk to people in the neighborhood and you may find there is more available land than you realized. This is also a great way to meet neighbors and start building community.

7. SOURCE YOUR SUPPLIES. Now that you are ready to get started, you will need to source things like good compost, seeds, plants, and maybe even equipment such as a rented tiller or tractor. This is a good time to find a local organic gardener or garden center that can help you refine your plan, source supplies, and even do some of the heavy lifting.

8. GET GROWING. Take a soil test and put your plan into action. Amend, till, plant, cultivate, and reap.

9. KEEP RECORDS. The last step in planning is keeping records of what happens in your garden. Though it can be a bit laborious to keep up with, this information will serve you well in the future. Some good things to track are planting dates, harvest dates, frost dates, air and soil temperatures, bug infestations, rainfall, and soil amendments.

10. HAVE FUN AND SHARE THE HARVEST. Growing food can be a serious business, and possibly a means of supporting yourself and your family, but it is always best done with a smile, so remember to have fun and share the love.

THE POWER OF VISIONING

"Visioning" is a process that enables you to think into the future and imagine what it is you want to create for yourself, your family, your farm or garden, your business, your community, and the world. Many companies these days include a visioning session with every client for a new project, and mine is no exception. Each consulting project I do starts with a type of exploration to gain a deeper understanding of the land, people, resources, and desired outcomes. The result is a report with project goals, opportunities and challenges, timeline, and action plan. This visioning document serves as the starting point for designing and programming for the project.

Planning and design takes everything into consideration to determine where things go on the land, how they connect to one another, and what kinds of experiences people will be able to have there. The goal with every project is to engage in the highest level of stewardship, share great-quality produce, and sustain the enterprise in perpetuity. No matter what the scale and scope of your project, a visioning session will help guide you through this journey. Encourage everyone involved to come into the visioning process with an open mind and positive attitude. This is the time to think outside the box and dream big; there are no limits to what you are allowed to imagine.

Start by setting some desired outcomes and then think about how you may be able to accomplish them. Determine the motivations and guiding principles that are driving the vision. Once you have a vision for where you want to go, taking steps and making decisions along the way will be much easier.

SIX STEPS FOR CULTIVATING A VISION

Step 1: Believe in yourself.

Step 2: Destroy any beliefs that are holding you back, especially ones that inhibit you from accomplishing step 1.

Step 3: Write out your vision. Be sure to do this from a place of confidence, humility, and positive intention.

Step 4: Let your vision percolate through every pore of your being and share it with those who will be a part of making this vision a reality.

Step 5: Write out your immediate goals with a timeline for accomplishing your vision.

Step 6: Get to work making your vision a reality. Remember to have fun and stay open-minded and grateful at every step along the way. Keep in mind that this is an ongoing process.

DEFINING YOUR INTENTIONS

As an urban farmer and educator, I get to go into many backyards and see what's growing, and sometimes what I see growing is frustration. This may be because of pests or soil imbalances that affect these gardens, and those are things that can be easily fixed. Sometimes, however, the frustration comes from a garden that doesn't quite meet the intentions of the gardener. More often than not, this happens when people don't take the time to create a vision and then formulate a realistic plan.

Before you plant a garden, you need to ask yourself a few questions to understand your intentions and thereby increase your satisfaction with your garden (or inspire you to start one):

1. WHY DO YOU REALLY WANT A GARDEN? If you can define this and keep it in mind, it will help you make decisions about your garden and connect you day-to-day with its bigger purpose in your life.

2. WHAT DO YOU WANT IT TO LOOK LIKE? Once you understand what your dream garden truly looks like, you can take specific steps to make it a reality.

3. HOW MUCH TIME AND MONEY DO YOU WANT TO SPEND ON IT? Making a realistic prediction about your resources will help you make decisions about the size and complexity of your garden.

★

TOOLS FOR PLANNING

Like the tools in any garden shed, our planning skills need to be sharpened now and again. To do this, I recommend mapping out your vision in your preferred medium. While I have a deep appreciation for primitive tools, I also have found modern technology to be very helpful in organizing my life. Here are some of the low-tech and high-tech tools I find useful for planning:

MEASURING TAPE AND WHEEL: It is very important to take measurements when planning a garden, for figuring out the size and quantity of raised beds, the amount of soil and fertilizer needed, the number of plants that can fit in an area, and the desired spacing in and between rows. For very general planning out in the field, I often use my step to measure 3-foot strides, but when getting into more detailed layouts, I use a measuring tape or measuring wheel. A tape is more than enough for small gardens, whereas a measuring wheel is more useful for larger plots or farms. Choose a measuring device that is a bright color so you don't lose it in the garden.

JOURNAL AND CROP PLAN: A garden journal is a great way to plan for the future and document the past. The more records you keep, the easier it will be to refer to lessons learned in the past so you can plan for success in the future. A fertility plan is a good way to monitor and enhance the health of your soil, and a simple chart or journal is a great way to keep track of soil amendments applied throughout each season. Evaluating how crops perform is very helpful for understanding the condition of the soil as well as how certain varieties fare in your climate and how they react to different types of amendments. I also like to keep track of planting dates, temperatures, and applications of fertilizer, compost, and pest controls.

MAPS: Maps of the property you are planning to farm or garden are extremely helpful and these days, thanks to the Internet, they are free and easy to find. By using Google Maps or Google Earth, for example, you can locate the site, zoom in, and print to scale. You can mark up digital or printed maps or use tracing paper to draft an overlay of ideas that come up in planning sessions. Be creative; tracing paper is cheap.

SOIL SURVEY REPORT: Thanks to the USDA's Natural Resources Conservation Service, every county in the United States has a soil survey report that shows what types of soils are in every part of the county. You can find your property and gather information on what types of soils are under your feet, general fertility, drainage, and more. This is best used for large-scale farm projects, but you never know what you might find, and it's free to look. Soil survey books are usually available at your local library or university extension office. The NRCS has also begun putting many online at http://websoilsurvey.nrcs.usda.gov/.

DRAFTING TOOLS: For sketching, I often rely on a good old-fashioned graph paper pad and pencil. Graph paper is great for backyard design, especially if you are breaking it down to the square foot. There are also some handy planning applications and programs online that allow you to plan your garden digitally. (Check out Mother Earth News Vegetable Garden Planner and GrowVeg.com as examples.)

When doing design work, I like to use drafting paper, an engineer or architect scale, and good-quality pens and markers. I use tracing paper placed over a map of the garden site and then use a pencil or fine-point marker to begin sketching. For beginners, I recommend using Col-Erase colored pencils, which come in lots of colors and erase easily. For pens and markers, I recommend Pentel sign pens, Pilot fine-liner pens, and AD markers.

ELECTRONIC PROGRAMS: Some applications for computers, tablets, and phone devices that have been very useful to me in various aspects of garden planning and design are:

- **SUNSEEKER—**maps the arc of sun exposure to help determine optimum garden and plant placement.
- **EVERNOTE—**organizes project notes by category and tags and is an heirloom in its own right as a brilliant new approach to creating and sharing notes.
- **SKITCH—**sketches on top of designs, photos, or maps with arrows, shapes, and notes for communicating visual ideas simply and quickly; also links to Evernote and email.
- **NOTABILITY—**using a stylus on an iPad, creates organizational charts, graphic notes, and diagrams.
- **OMNIFOCUS—**for the big picture; organizes projects by category for tracking and productive management of multiple tasks.
- **DROPBOX—**for storing and sharing designs, photos, and more.

PLANNING GARDENS FOR DIFFERENT VENUES

Gardens can range in size and scope from a simple planter on the patio to a full-blown farm-based neighborhood with community gardens and edible streetscapes. The following is a list of places—from your backyard to the back forty—where gardens can be conduits and catalysts not only for food, but also for education, recreation, entrepreneurship, healing, and building community.

HOMES: There are millions of acres of backyards, front yards, and side yards that could be growing food for people and wildlife. Whether you just want a small herb planter on the windowsill or have visions of turning your yard into a mini farm, the home is a perfect place to grow. Instead of a big lawn that sucks water and feeds no one, I recommend edible and native landscaping, raised-bed veggie gardens, rain barrels, a compost bin, and maybe even chickens and bees (if your county or homeowners' association allows them).

SCHOOLS: Incorporating gardens into schools at all levels—from preschool to college—can help young people establish better eating habits, learn how to grow their own food, and become more conscious and skilled stewards. Having a garden at school is a great opportunity to apply science, nutrition, math, and social studies lessons, to name a few. It is also an opportunity for parents to get involved with their school PTA by helping out and becoming part of the community that can spring from a shared garden. At the college level, farms become a research laboratory, interdisciplinary classroom, and student hangout.

RESTAURANTS: Having fresh herbs, produce, and flowers right outside the restaurant door is a chef's dream come true. Chefs understand the difference in quality between something picked fresh and something shipped across the country. More and more restaurants are squeezing gardens into their landscape or rooftops, or even setting up their own farms nearby.

SPAS AND RESORTS: Most resorts and spas these days are using natural ingredients in both meals and treatments. Some are even taking it one step further, with farm-to-table or farm-to-spa gardens on site. Herbs and flowers are perfect for a spa garden, as they are beautiful and relatively easy to maintain. Resorts can benefit greatly by incorporating a chef's garden or produce from a small farm into their restaurant menus and catering, while also serving as a venue for events and workshops.

HEALING PLACES: Gardens can provide a healing space as well as nutritious food and medicinal herbs. Gardens can be a peaceful place to rest, reflect, and restore. Horticultural therapy is a growing profession, and there are many horticultural therapy activities for people with a wide variety of issues—from basic stress and anxiety to more serious illnesses. There are also social, physical, cognitive, and psychological benefits that can be realized in a garden.

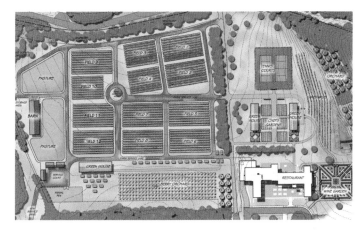

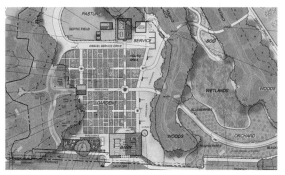

FAITH-BASED INSTITUTIONS: Stewardship is a virtue that has roots in many different faiths. It is a shared belief that we are responsible for taking care of the earth and leaving it in better shape than we found it. Unfortunately, we are not living up to that goal, and many faith-based groups are taking action to raise awareness within their communities about steps that can be taken to help all of us become better citizen farmers. One of the ways this is implemented is through gardening. More and more religious schools, summer camps, community centers, and places of worship are incorporating gardens for education and charity. In addition to gardens, many of these organizations are promoting and facilitating CSA farm pickups within their congregations and communities.

CORPORATE VENUES: The most common corporate gardens, though they are not as common as I wish they were, are at large corporate campuses where land is plentiful and there are enough employees who wish to participate. If I were working at an office all day, I would love to be able to take my lunch breaks in a garden and put some fresh greens and tomatoes on my sandwich. At my store in Atlanta, the garden-center and car-wash staff work together tending about thirty garden beds and the produce is shared equally. Everyone also helps care for the chickens and bees and shares in the harvest of eggs and honey. There is lots of potential for integrating gardens into businesses and corporate campuses across the country to improve the quality of food and corporate culture. It is a great way for companies to engage their employees in a team effort as well as to promote health, sustainability, and giving back to the community by donating food to local food banks.

HISTORIC SITES: Throughout the country there are farms that have been or should be protected as national historic sites. These are wonderful opportunities to preserve the agricultural heritage of an area and teach children about the way people used to live and farm. These are also often some of the last undeveloped green spaces in sprawling suburbs where farming can still take place. I have worked on a few very interesting historic farm projects, such as the Trustees Garden in Savannah and Hyde Farm in Marietta, both of which provide opportunities to celebrate history and bring agriculture into an urban context, where it is sorely needed.

NEIGHBORHOODS: I believe the future is where a farm or community garden is at the heart of a neighborhood of homes. This way people are connected to their food, and the carbon footprint that comes with shipping fresh products all over the place is greatly reduced.

One could say agriculture is the "new golf," meaning that instead of building neighborhoods around golf courses, developers are building them around organic farms. This still provides open green space, but it also cultivates food and community rather than consuming massive amounts of water (and in most cases toxic chemicals) to play a sport. Now, I love sports and golf is a great game, but I believe there are more than enough golf courses and nowhere near enough neighborhood farms!

CSA AND MARKET FARMS: One of the best modern solutions to the challenge of bringing back the small family farm is Community Supported Agriculture (CSA). This is a way for farmers and consumers to have a direct relationship with each other, in which they both share in the risks and rewards that come with growing local organic food. CSA members pay for a season's worth of produce in advance, so that the farmer has seed capital with which to get fields prepared, seeds started, and help hired. The return on investment for the members is a weekly share of the harvest—fresh-picked organic produce with no middleman.

This makes it more affordable for the farmer to grow and for the consumer to afford local organic products. It also builds community and enables farmers to develop other aspects of their businesses, such as selling to restaurants, farmers' markets, and wholesalers. I have been a CSA farmer for many years, and while it is very hard work, it is the most fulfilling business relationship I have ever been in. I highly recommend that you check out localharvest.org to find a CSA near you.

HOW TO PLAN A COMMUNITY GARDEN

Community gardens have become the new hangout at parks, schools, places of worship, previously empty lots, and any other little swath of land that can be transformed into a vibrant spot where both food and community are cultivated. Many community garden leaders intend their locations to be welcome havens for all, but achieving that can sometimes be challenging. Here are some ways to ensure that all members of your community can participate in the garden and enjoy the fruits of their labor:

1. MAKE SURE YOUR LOCATION IS EASILY ACCESSIBLE. Can you see the garden from the street, or is it blocked by weeds and debris? Is there available parking, including a bike rack? Do you have signs letting people know that it is a community garden and all are welcome to visit? If not, these are all easy, inexpensive fixes that might just mean clearing out and designating a little extra space to make your members and visitors feel more welcome.

2. ACCOMMODATE THOSE WITH DISABILITIES. If you haven't thought about accessibility for people with disabilities, you'll think of it pretty quickly the first time someone with a cane, walker, or wheelchair can't get

up your wood-chipped path. This is where a partially or fully paved path and an extra-high raised bed that a wheelchair can slide under or beside come in handy. Many community gardens are incorporating these elements right from the get-go. Other amenities that take into consideration a wide range of abilities include shaded areas, benches, convenient restrooms, and sensory integration elements, such as herb gardens, wind chimes, art elements, water elements, food sampling areas, and other features where all five senses are awakened.

3. MAINTAIN AN OPEN DOOR (GATE) POLICY. There comes a time, usually around the first tomato-snatching, when community garden leaders decide to lock their gardens to reduce the chances of theft. Best practices nationwide show proven strategies to achieve this goal without locking out the public. These include encouraging members to personalize their beds and harvest frequently to show that the food grown is valued, dedicating excess produce to a local food pantry to show commitment to those in need, and offering a "thieves' bed" where anyone can pick whatever's ready to be harvested. Also, keep finding ways to involve more factions of the community in the garden, and you will increase the number of people who feel ownership and pride in it, and will therefore help protect it. Buddying up with your local police department is always a good idea—you could even offer your precinct a few free plots in exchange for their support.

4. PROVIDE A WELCOMING ATMOSPHERE. Community garden leaders can set a good example for their members by always greeting visitors to the garden, engaging them in conversation, and inviting them to help or browse. People who linger, laugh, and learn together create a positive environment that is attractive to others. Some ways to know if you're a quart low on this attribute are: (a) if your garden members are simply running in, watering, and leaving; (b) if your garden members don't know one another's names; (c) if you don't have a steady stream of nonmembers coming by to check out the garden; and, of course, (d) if going to the garden is simply not fun.

5. IF YOU WANT TO GROW FOOD FOR SALE, CHECK THE RULES. You need to determine whether or not commercial activity is allowed, as well as how you can accommodate larger equipment, food storage, accessory structures, and a larger compost operation than a home or community garden might require. Contact your city hall for exact details about local ordinances that affect growing food where you live, so you can be ready to dig in when it's time for spring planting.

GET INVOLVED IN A COMMUNITY GARDEN—
WITHOUT GARDENING!

The beauty of a community garden is that anyone can get involved, even those with physical limitations, brown thumbs, and lack of time or funds.

Maybe you just want to stroll through the garden and enjoy it during your morning walk. Most gardens are very welcoming to visitors and are happy you are appreciating the oasis they created. Say hello, compliment their efforts, leave your dog outside the garden, and take any trash you have with you. If you feel like it, it's always nice to offer to help with something, however briefly.

If you want to get more involved but don't necessarily want to become a member or take care of your own plot, ask if there are any teams you could join occasionally, perhaps for making compost piles, or growing food for those in need.

Let garden leaders know about any special skills you may be willing to offer. Community gardens and farms often need help with marketing, fund-raising, grant writing, legal issues, and accounting. Also, there always seems to be something that needs to be built or repaired. An engineering type may be just the person to get that aquaponics or rain-harvesting system up and running, or to help build a hoop house for winter growing. Looking for a position on a board of directors? You may be surprised how easily you could find yourself in that role! The experience you gain and share in these volunteer capacities can help you in your professional life by sharpening your skills and broadening your network.

Who knows? You may be the perfect person to write the garden blog, serve as the resident photographer, or establish relationships with local restaurants to compost their food waste.

PLANNING FOR ANIMALS IN THE GARDEN

Organic gardening is more than simply growing healthy food. It also means creating your own personal eco-system where nature works with you and you work with nature. Barriers and deterrents can help you reduce theft by critters, but giving nature time to work things out can help, too. For instance, an abundance of chipmunks attracts majestic red-tailed hawks that can help keep supply and demand in balance. And never underestimate the value of a good hunting cat or guard dog.

The fringe benefits of including animals in your garden plan are endless. What's nicer than getting up in the morning and going outside for freshly laid eggs for your garden-to-table breakfast? Or how about some

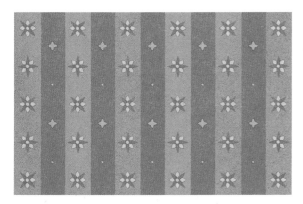

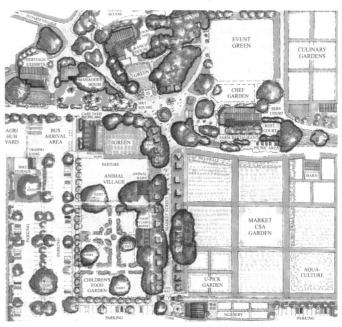

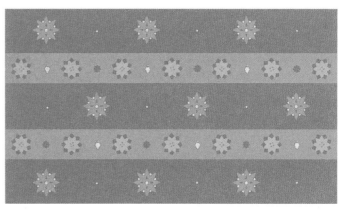

homemade goat cheese from goats you milked yourself, or tilapia baked in parchment from the aquaponic system you set up in your mini greenhouse? Yes, this is all possible in a small suburban or even urban garden setting.

If you are interested in adding animals to your home or community garden, here are some tips to get you started:

1. CHECK LOCAL ZONING ORDINANCES AND THE COVENANTS OF YOUR HOMEOWNERS' ASSOCIATION. This is important. You don't want to go to the expense and trouble of setting up a habitat for your chickens or goats and then find out that they are not allowed, or that there are setback requirements for a chicken coop or rabbit hutch that you didn't follow. Be a good neighbor about this, and you are more likely to have support in your community.

2. RESEARCH BEST PRACTICES FOR THE CARE OF THE ANIMAL SPECIES YOU HAVE IN MIND. A withering tomato plant is one thing; a suffering animal is something entirely different, and a little education can go a long way toward raising healthy, happy animals. Read books, take a class, take a local chicken coop tour, talk to friends and neighbors, join a group, help out on a small farm nearby, or join a community garden beekeeping team. You may find that the responsibility of having animals is greater than you realized, and that you'd prefer to support a local farmer by purchasing eggs, cheese, and honey, rather than producing them in your own garden.

3. CONSIDER CREATING HABITATS FOR WILD ANIMALS. If you decide not to raise animals, or if your city doesn't allow it, don't give up. You can enjoy a wide range of animal species by creating a welcoming environment in your garden. Attract pollinators like bees and butterflies by planting flowers, herbs, and flowering vegetables that they like. Encourage bats (which eat lots of mosquitoes and other bugs) by adding a bat box. Invite birds into your yard with bird feeders and by growing towering grains like amaranth and sorghum. And even make bunnies feel welcome by planting crimson clover for them (which they prefer over almost everything else). Let some part of your garden stay a little wild so that animals have a covered area for breeding. Add birdbaths and fill them often. Put some rocks around where lizards and frogs can hide. Sit and enjoy your diverse ecosystem, and you may find that this easy way to have animals in your garden is more than enough for you.

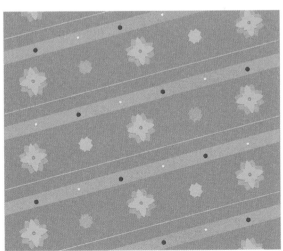

★ TOP 10 ★

EDIBLE AND MEDICINAL PERENNIALS
TO REPLACE LANDSCAPE ORNAMENTALS

This is for those of you who are tired of pretty, but pretty useless, landscape plants taking up valuable space in your yard. These are great edible perennials that can provide food and medicine year after year.

1. **BLUEBERRIES:** Possibly the best edible perennials you can add to your landscape. They make excellent hedges and landscaping around your home. Blueberries are delicious, nutritious, and pretty low-maintenance. I recommend planting a few different varieties both for pollination and to extend your harvest, as there are early-, mid-, and late-season varieties. Blueberries do best with full sun but will also tolerate partial shade.

2. **RHUBARB:** This perennial vegetable has beautiful red stalks and large crinkly leaves. While it is a wonderful addition to cobblers and pies, beware: Its leaves are poisonous, so only eat the red stalks. Grow rhubarb in full sun unless you are in a very hot, dry climate; then plant where it can get some afternoon shade. Rhubarb grows 2 to 4 feet tall. Plant from crowns in the early spring or fall.

3. **ECHINACEA:** One of the best-known medicinal plants, this is a gorgeous perennial that will keep your immune system strong in addition to adding color to your landscape. Plant in early spring or fall, from seed or transplants; plant in full sun with well-drained soil. Echinacea seed is also easy to save by harvesting the stems and hanging them to dry. Wait to dig up roots for making medicine until plants are three years old.

4. **SORREL:** This is a crowd pleaser, especially among kids. Easy to grow from seed, sorrel is a perennial herb that is rich in vitamins and minerals and has astringent properties that help clean the blood. Cut back stems when they start to go to seed and harvest leaves as needed. They grow 16 to 24 inches tall and like full sun or partial shade.

5. **MULTIPLIER ONIONS:** There are two main types of multiplier onions: a top-setting multiplier onion (also known as an Egyptian or walking onion), and a bulbing multiplier onion (also known as a potato onion). All of these edible perennials can be planted from bulbs in fall; once established, they will multiply, hence their name. They do best in full sun or partial shade.

6. **CULINARY HERBS:** Perennial culinary herbs like chives, oregano, rosemary, sage, and thyme are beautiful additions to any landscape. The creeping varieties are great for rock walls and edges of containers, while the more upright types are perfect for planting in beds and as centerpieces in containers. Culinary herbs typically thrive in full sun to partial shade and do best in well-drained soil.

7. **GRAPES:** A wonderful vining perennial that can provide shade, beauty, food, and wine. They are best planted in winter, when the vines are dormant. Plant in well-drained soil in full sun, with an arbor or trellis to climb. Grapes are perfect for growing over a garden arbor, fence, or pergola. Prune, fertilize, and pick to keep your vines healthy and under control.

8. **FRUIT TREES:** Planting fruit trees, especially dwarf varieties, can transform your landscape from purely ornamental to something that produces food with minimal effort. These are best planted in winter, when they are dormant; they should be kept pruned and as disease- and pest-free as possible. To get started, I recommend planting whatever fruit does best in your region, with the least amount of care and expertise. You can always advance to more finicky trees later. Fruit trees generally need full sun.

9. **NUT TREES:** Nut trees are great food-producing shade trees that can provide a substantial source of protein for your family as well as a crop of leaves for the garden. Check with your local university extension or fruit tree association to find out which specific varieties do best in your region. Most nut trees do best in full sun.

10. **VINE BERRIES:** A hedge of fruit-bearing vines can be a wonderful property border, feature in the landscape, or means of hiding a utility box or other eyesore. The native berries have thorns, which can be useful to deter animals (or people). There are also thornless varieties of blackberries, raspberries, and other vining berries that can be grown without danger of kids or pets getting caught in the briars. Check with local resources for the best varieties in your area. They prefer full sun but will tolerate partial shade, especially in the afternoon.

★ TOP 10 ★

ANNUAL CROPS FOR EDIBLE LANDSCAPING

This is for those of you who want your yard to look beautiful but want to eat it, too. These are all annuals that can be planted to provide your front, back, and side yards with seasonal color, food, and wildlife habitat.

1. **KALE:** Great cool-season annual that adds beauty and food to your landscape. My favorite varieties are Toscano dinosaur kale, Redbor, and Red Russian.

2. **SWISS CHARD:** A beautiful plant for any landscape; it can withstand some shade and in many climates will live longer than most green, leafy vegetables. Easy to grow from seed or transplants. It is best to pick off any dead or dying leaves, as chard is prone to pests and does best when kept clean and weed-free.

3. **NASTURTIUMS:** These beautiful and edible flowers (and leaves) are great for growing up fences, arbors, and trellises. They like cool weather, so plant in early spring and late summer for fall blooms. In some cooler climates, they can be grown all summer long. I have had fun growing these up sunflower stalks in the landscape.

4. **SUNFLOWERS:** Easy and fast to grow from seed, these add beauty and bees to the landscape. They come in dozens of sizes and colors, so there is lots of room for creativity here. They also make nice trellises for other vining plants, like nasturtiums, pole beans, and cucumbers.

5. **EDIBLE VIOLAS AND PANSIES:** These are nice border plants to put around the mailbox and landscape beds in spring and fall. Some will tolerate hot weather if they have some mid-afternoon shade. Both have edible flowers and a few different colors to choose from.

6. **OKRA:** A heat-loving annual that has gorgeous flowers, leaves, and fruit. Okra plants can reach heights of 10 feet if you let them, which can be beautiful in the landscape. You can also cut them back at 2 to 4 feet to get a bushier plant. The pods of okra are best harvested for food every few days, but if you miss a few, they add interest to the landscape. There are red varieties such as Red Burgundy and Hill Country Red, which add wonderful color as well as height and texture.

7. **BASIL:** One of the most beautiful edible ornamentals; grown for its fragrance, color, taste, and all the insects it attracts. It is easy to grow from seed or transplants, and should be planted after the last frost. If kept free from disease and bugs, basil can grow all summer long. As a landscape plant, I don't mind if it flowers, but if you are growing it for its fresh leaves or to make pesto, I recommend harvesting often to prevent bolting (going to seed). My favorite varieties are Thai, purple ruffles, African, lemon, Genovese sweet, and holy basil.

8. **LETTUCE:** Lettuce is a wonderful addition to the annual landscape. There are many varieties that will add color and texture to the landscape, as well as being delicious. Lettuce can tolerate some shade, and actually prefers it in the summer months. Plant slow-bolting varieties to get the most out of a landscape lettuce.

9. **SHISO:** A lesser-known herb that is easy to grow and resembles basil. It is used often in Japanese cooking—you've very likely seen it on a sushi platter. There are green and red shiso; both are beautiful, and once established, they will reseed easily. It is sometimes considered a weed, so keep an eye on it if it starts to pop up everywhere, but don't be discouraged—it is very easy to remove.

10. **SCARLET RUNNER BEANS:** A beautiful vining bean with seeds and roots that can be eaten, though it is grown more for its beauty than as food. It is best to pick the bean pods regularly to encourage the blooms. It is a favorite with hummingbirds, and if cared for, it will bloom all summer long. Plant after any danger of frost.

"To forget how to dig the earth and to tend the soil is to forget ourselves."
Mohandas K. Gandhi

3.

TILLING = INITIATIVE

I must have a special kind of karma that attracts me to areas with some of the worst soil imaginable for planting an organic farm.

When I arrived at Log Cabin Ranch, the youth prison farm where I worked in California, I pulled a tarp off the weed-choked ground to find snakes nesting in large dried cracks of the rock-hard soil, impenetrable even with a pickax. When starting the farm at Serenbe, I had to unearth mattresses and old tires before I could even begin to plow the eroded red-clay soil. While laying the groundwork for Hampton Island resort near Savannah, I inherited a field of pond muck—also called "gumbo"—that had a soil test of a very acidic 3.5 pH and a handwritten note from the lab that read: "Hope you're not planning to farm this." At the historic Trustees Garden in Savannah, Georgia, which had been abandoned for more than two hundred years, my challenge was to grow food atop what had been a grass parking lot where the soil had become extremely compacted by large trucks that used it daily as a turnaround.

All of these seemingly hopeless plots, I am proud to say, eventually came back to life and thrived. Getting there took years of patience and buckets of sweat, but I found that the harder I had to work at it, the greater the rewards—not only in terms of an abundant harvest, but also in how it made me feel physically, mentally, and even spiritually.

In the previous chapter, my goal was to get your creative juices flowing to plan the garden of your dreams.

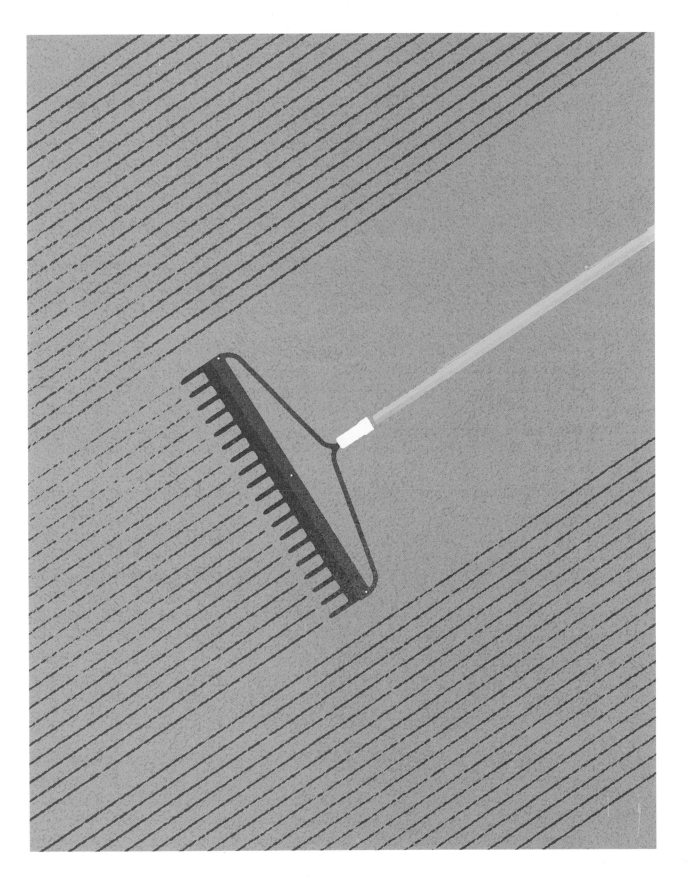

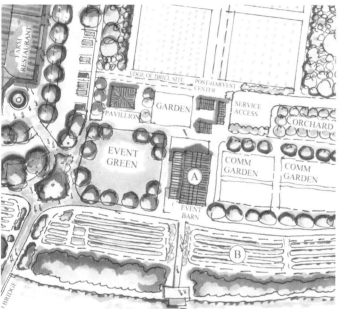

EDGE OF DRILL SITE — POST-HARVEST CENTER

FARM RESTAURANT

PAVILLION

GARDEN

SERVICE ACCESS

ORCHARD

EVENT GREEN

A

COMM GARDEN

COMM GARDEN

EVENT BARN

B

BRIDGE

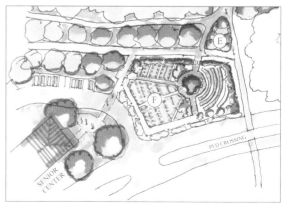

SENIOR CENTER

E

F

PED CROSSING

Now it's time to get your hands dirty. In the pages that follow, I hope to equip you with the skills to build a healthy foundation for your first planting and to maintain and enhance your soil's fertility over time. I will share some tools and tips to set you up for success, and I hope after reading this chapter you will feel inspired to grab a digging fork, chop saw, or tiller and get growing.

GROW FOR IT

Here are four easy steps to get you growing:

1. CHOOSE YOUR LOCATION. As part of your garden planning, assess your property for sun, slope, soil quality, and water access. A vegetable garden requires at least six hours of sun a day and should be situated to take advantage of it (south-facing seems to work best). If it is winter, picture the trees with leaves on them to be sure they won't shade the garden site in summer. Take a "sun" day: Spend a whole day in the garden watching and taking notes as to when there is sun or shade in different areas. The sunniest areas are going to be those facing south, as the arc of the sun moves from east to west, higher in the sky in summer and lower in winter. The better you understand your site, the more you will be able to find microclimates where certain crops will thrive.

Microclimates are "atmosphere zones" that are different from the surrounding area. They can be the size of a small raised bed, or acres wide. They are created by a large combination of factors, from varied topography including hills, water, and trees to urban elements such as tall buildings, brick walls, and concrete. Gardens that face different directions can have different microclimates, so it might be worth your while to grow on all sides of your house. One way to assess your different microclimates is to plant the same type of plant on various parts of your property and then observe their growth over time.

As you consider locations around your home, don't discount the sides and front of your property. If you live in a condominium or apartment, ask your management company if there is a place you might be able to start a garden. If not, you may be able to plant one on your patio or balcony, or you could simply put some herb planters in your kitchen window or consider one of the indoor hydroponic gardening systems available now.

2. GET YOUR SOIL TESTED. Soil is the source of our sustenance, and when nurtured properly it can nurture us back. Soil is a living organism that needs to drink, eat, and breathe. Well-balanced soil helps plants grow strong, enabling them to fight off pests and diseases so you don't have to spend time and money doing that for them.

If you live in an urban or suburban environment, chances are your native topsoil has long since been removed. Whether you are using native soil or transferred soil, the first thing I recommend is taking a soil sample. This is an easy and inexpensive exercise that will start you off on the right foot. A soil test can show you what your soil needs to achieve a balanced chemistry, such as pH, calcium and magnesium ratios, and trace minerals. Some tests can also show you what types of living organisms are in your soil such as beneficial fungi, bacteria, and protozoa. A heavy-metals test will also let you know whether the soil contains lead or any industrial toxins that you simply don't want in your food. This information will help you steward your soil to an optimum state for producing safe, healthy, nutrient-rich food and medicine.

You'll need to contact your county's agricultural extension office or a private soil testing laboratory for specific directions on how to submit a soil sample for testing, but the gathering of the sample will always follow this basic procedure: Using a soil probe, trowel, or shovel, dig a hole 6 to 8 inches deep and put the soil in a clean bucket. Take samples every 5 to 10 feet for a home garden and 25 to 50 feet for an acre or more. Mix all samples together for each chosen area and fill a bag with about 1 cup of the well-mixed soil sample. Mail this to your county extension office or soil-testing laboratory and ask to have your results converted to organic recommendations.

It's best to get your soil tested at least a month before your first planting season, and then annually after that. Why? Because many organic amendments like lime will take time to become available in the soil. The more balance and biology you can cultivate in the soil, the more available nutrients become.

3. INSTALL THE BEDS. Determine the size and shape of your garden and then build your planting beds (or you can purchase them premade). These can be mounds of soil right on the ground or raised beds made out of untreated wood, cinder blocks, stone, or whatever nontoxic material you have available or can afford to purchase. If you are using pots, be sure they have drainage holes. You may need to drill holes in pots you purchase.

Although it may be tempting to repurpose railroad ties and other free wood, think twice. Make sure to use untreated wood so that you are not introducing unwanted toxic chemicals into your garden. When you choose untreated wood, give some thought to longevity when you evaluate cost. Pine is the least expensive, but you'll need to replace it every few years and its appearance deteriorates after a while, so if looks and longevity count, this is not your best choice. Cedar is more expensive but it can last for more than ten years, and it weathers nicely. We build most of our raised beds out of 2-inch-thick, 10-inch-wide, untreated FSC certified Western Red Cedar. We like using cedar because it is naturally rot-resistant and has a lower carbon footprint than most other wood options. Also consider the source by choosing wood harvested from sustainably managed forests, preferably in your region to reduce transportation impact.

Other materials such as repurposed cinder blocks make nice, affordable garden beds but they do dry out quickly. Old tires are not recommended, as they can leach toxins.

Raised beds offer some unique benefits: The wood frames keep your soil from eroding, the soil heats up earlier in the spring and stays warmer later in the fall, the soil drains well after rains, and crops in raised beds can be protected from burrowing animals by attaching some chicken wire to the bottom. They are also easy to install (and to move) and are great for areas with poor or rocky soil. They tend to have fewer weeds and higher yields per square foot, because you can fill them with premium weed-free soil that is balanced, rich, and well drained.

A 10-inch-deep bed is what we typically recommend, as that is adequate for growing almost anything, especially if the bottom of the bed is open to the native soil. We also build beds as tall as 3 feet to eliminate bending and make the garden easily accessible by wheelchair.

4. ADD AND AMEND YOUR SOIL. The ideal growing medium for plants and vegetables is loam. It is a combination of soil; organic matter, such as compost; nutrients, including potassium, nitrogen, phosphorus, and trace minerals; and air. (If you have good, healthy, nontoxic soil, the earthworms this attracts will burrow holes that give your soil air.) Follow your soil test results to create this perfect mixture, or consider buying a premixed version. An easy-to-do pH test, which you can buy at a garden center, will let you know the pH level of your soil. Most veggies like the pH to be a little below 7. You can raise the pH by adding lime or bring it down with gypsum or coffee grounds.

Add fertilizer and compost before planting, then feed a little bit to the crop every few weeks, or once a month, based on how it is performing. I recommend adding significantly more compost in your first season to provide the soil with a boost of organic matter. To give you an example, organic farmers often add as much as 50 tons per acre in their first year and 10 tons per acre every year thereafter.

If your city or county offers municipal compost to gardeners (usually created from yard debris picked up curbside throughout the city), be sure it does not contain sewage sludge or persistent herbicides. Start your own composting system as soon as possible so that you can have a trusted, free source of compost, and check out reliable local suppliers for additional soil amendments. Ask around at community gardens and farmers markets—they often know all the best dirt in town.

★

TOOLS FOR TILLING

GARDEN TROWEL: Possibly the most practical tool for the garden is a basic trowel. It is so important and symbolic that I chose to use it as my logo. There are lots of cheap trowels out there, but they are unlikely to be very comfortable or long lasting. A high-quality hand-forged stainless steel trowel with a wood handle will cost from $20 to $40. There are many types of trowels for everything from digging and scooping to transplanting and potting, so make sure to pick the one that fits your task.

DIGGING FORK: A high-quality digging fork is possibly the most important tool for a home gardener. I use a digging fork for creating new beds and for aerating existing beds between crops, as well as for turning in compost, cover crops, and fertilizers. Invest in something with heavy-duty tines and a solid handle. A decent digging fork can be bought for $25 to $50, while higher-quality forks go for $75 to $150.

GARDEN SPADE: An all-purpose garden spade is something you will use often for planting, transplanting, dividing perennials, and shoveling compost. While you can pick up something inexpensive at a hardware store that may last you a year or two, I recommend investing in one that you'll love to work with and that will last for decades. A hand-forged spade with a D-handle will cost from $50 to $100.

EYE HOE: A good hoe is hard to find. While most hoes are for weeding, some work well for digging—like a good eye hoe. I use eye hoes for everything from leveling beds and breaking up compacted soil to turning in cover crops. A heavy-duty eye hoe will cost from $25 to $60.

CULTIVATOR: There are many types of hand cultivators, such as the classic three-, four-, or five-tined rake style, single-hook type, and solid Korean plow cultivators. These are very handy for hard-to-get weeds growing between and around crops. A good-quality hand cultivator will cost about $20, and long-handled versions two or three times that.

GARDEN RAKE: A sturdy garden rake is an essential tool for any vegetable gardener. I use one to level the surface of the soil before seeding, for raking out weeds, for gently covering broadcasted seeds, and for cleaning out beds of rocks and other debris. You can buy a cheap rake at a hardware store, but it's likely to get wobbly on you after a few good tugs. I like a long-handled rake with between eight and twelve sharp stainless-steel tines. A sturdy garden rake like this will cost between $50 and $100.

BROAD FORK: I first learned about broad forks when I was at a biodynamics workshop in 1995, learning how to dance down the edge of a bed while gracefully double digging and breaking up clumps. Broad forks are extremely effective for aerating beds. These niche tools are not cheap; they range from $150 to $250.

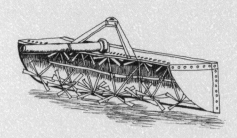

ROTOTILLER: If you have outgrown your digging fork and want to till up the front yard, you may be in the market for a good rototiller. It could be a bit much for you to invest in one alone, so consider going in on one with neighbors or community gardeners. The problems with rototillers are that they can harden the soil, and they are dangerous—they should be operated with closed-toe shoes and far away from children and pets. The benefit of a rototiller is that you can turn a lot of ground using much less effort than with digging forks and hoes. The only tillers that you will typically find at a garden center or hardware store are very basic models that run between $400 and $900. While these may do the trick for your small-scale garden, I recommend going with something a bit heavier duty, like a BCS or Grillo. A six- to eight-horsepower model of either should be more than adequate for your backyard. Models in this horsepower with a tiller attachment range from $2,000 to $4,000.

If you are planning on working an acre or more, I would suggest bumping up to something between ten and fourteen horsepower, which will cost more in the $4,000-to-$6,000 range. These are seriously powerful tools that can pull all kinds of implements besides tillers, such as plows, mowers, and even a rotary spader. Additional implements range from $200 to $2,000. You will be amazed at how quickly these can pay for themselves if you are looking to sell what you grow.

WOODWORKING TOOLS: If you decide to go with raised beds instead of in-ground beds, you may need some basic woodworking tools. The bare essentials include a chop saw and power drill with the right blades and bits for your materials.

TRACTORS AND IMPLEMENTS: I love old tractors and spent my first few farming years driving a 1940s Oliver and 1950s Ford. When I bought my farm, I decided to go with a much newer Kubota, as I did not have much mechanical experience to handle the maintenence. When it comes to small tractors, I am pretty partial to Kubotas, even though I have had positive experiences with John Deeres and Masseys over the years. My favorite combination is a forty- to fifty-horsepower Kubota with a Falc or Tortella spader. A spader combines the work of a plow and tiller, but does both better. A tiller cuts and compacts the soil, often creating a hardpan, whereas a spader digs and aerates the soil, breaking up the hardpan.

TO DIG OR NOT TO DIG

Here are a few alternative methods to building raised beds:

NO-TILL: There is a whole body of evidence that shows that the less you disturb the soil, the more you preserve the valuable microbial life happening below ground, and the more beneficial this is to your growing efforts. If you don't have nutrient-rich soil naturally in the place you want to grow, you can create a mounded bed with an organic planting mix and then choose to practice the no-till method after that.

An effective way to create a no-till garden from scratch is to make a "lasagna garden" in the fall that decomposes all winter and is ready for spring planting. Start with cardboard on the ground where you want to build a garden. Add a layer of compost, then layer with browns (leaves, shredded newspaper), then greens (grass clippings, garden trimmings), then browns, then greens, and so on, until your pile is about 2 feet tall. This pile will break down and shrink a great deal over time, and will be crumbly and ready for planting, without any tilling, in a few months.

DOUBLE-DIG: This method is particularly useful if you have compacted soil and is only for those looking to get a really good workout! It requires digging down about 12 inches, then digging another 12 inches. Use a wheelbarrow to move the first batch to the end of the row, and then work your way down, transferring each 12 inches of topsoil and 12 inches of subsoil into the trench you just finished digging. You end up with a planting bed that is fluffy and loose, so that plant roots can permeate deeply. This enables you to grow more in less space since the roots go down instead of out. Add organic compost and fertilizer to increase the fertility of the soil. You can do this with or without a raised bed frame; using one will maximize the growing potential of each square foot and is a great strategy for small-space gardening.

HUGELKULTUR: This is a German growing method in which you bury large branches and cover them with rich soil and compost, and then plant above them. The wood retains water, thereby significantly reducing your watering needs, and it decomposes slowly over time, continually fertilizing your garden. This is a particularly good option in areas that are prone to drought (see page 35 for more details).

CONTAINER GARDENS: If you don't have a big yard or just want to start small, I recommend a few pots, urban planters, or raised beds on legs. Start with some potted herbs like parsley, thyme, and oregano, build up some confidence, then graduate to growing some easy veggies like lettuce and kale. You will quickly discover how easy and fun it is, and before you know it you will be plowing up your yard to grow more food.

VERTICAL GARDENING: Vertical gardening offers terrific solutions for the home gardener who lacks space, loves to experiment, may have physical limitations that make crouching and digging difficult, or wants to incorporate an eye-catching design element into his or her home decor. Here are some ways you can take your healthy, homegrown food harvests to whole new heights:

- **TRELLISES AND ARBORS:** If you already have a productive outdoor garden, you may want to double your growing potential by adding vertical structures such as trellises and arbors. You can then grow vining perennials up them, such as grapes, or seasonal favorites such as pole beans and cucumbers. Place these structures on the north side of your garden so they don't block sun exposure for your other crops. If you build them big and sturdy, you can grow heavy varieties like gourds, as they will shade crops underneath that like it a bit cooler. Building or buying these types of structures is a good winter project.

- **WINDOW GARDENS:** If you already have sun pouring in your kitchen (or any other room), you are halfway to growing year-round. All you need is a system that uses that abundant sunlight to deliver fresh lettuces, herbs, and more in a neat, handy way. Vertical growing systems that attach to windows achieve this objective efficiently while also adding a nice decorative touch.

- **LIVING WALLS:** Take the whole vertical-window-garden concept a step further and include a fully automated system that can be adapted for either indoor or outdoor use. There are some excellent but expensive systems available for all kinds of applications. One very simple method is to connect several gutters horizontally to the wall of a garage or the side of a house, fill them with soil or a hydroponic watering system, and plant them with shallow-rooting veggies and herbs. These are a great way to turn unused space on the sides of your buildings into edible walls; they also serve as striking accents in a home or restaurant, or as elegant property dividers or patio features.

HEALTHY MULTITASKING:
GET YOUR EXERCISE WHILE GARDENING

Do you laugh out loud when people say they get their exercise by gardening? If so, you've probably never tended a community garden or volunteered at an organic farm where you may haul five hundred pounds of cow manure or plant thirty trees! Even putting a few wheelbarrowfuls of fresh wood chips around a raised garden bed may have you feeling sore the next day. Here's how to make it safe and effective:

1. ASSESS YOUR CURRENT FITNESS LEVEL. See your doctor if you have any concerns. Gardening involves walking, bending, lifting, twisting, crouching, and even heart-pounding cardio.

2. WEAR APPROPRIATE CLOTHES. Dress in layers when it's colder out, and no matter what the season, wear appropriate sun protection. Be extra conscientious about footwear—closed-toe shoes are best. When you're out in the garden, you're around sharp tools and, occasionally, an unexpected mound of fire ants.

3. WARM UP YOUR MUSCLES: DON'T JUST GRAB THAT PITCHFORK AND START TOSSING THE COMPOST PILE. Walk a bit. Do a few stretches. Loosen up the joints. Try a moving meditation like yoga or tai chi. Then establish a comfortable pace with frequent breaks to catch your breath and have a sip of water. Switch sides when doing a repetitive garden task. And don't try to be a hero—do what feels comfortable.

4. ESTABLISH A GOAL. If fitness is an important reason you're gardening, try to attain your objectives each time you garden. Perhaps you want to spread wood chips at the community garden for half an hour, or spend two hours building a fence. Voice your goal to others so that they don't distract you with conversation or brownies at the picnic table (at least not until later).

5. TAKE A LEADERSHIP POSITION. If you want to be sure you won't slack off on your fitness routine, lead a community garden team that requires regular action, like the compost team or the grounds committee. You might also offer to mow the lawn or pick up the giant tubs of kitchen scraps from local restaurants. And keep an eye out for opportunities to get your heart pumping at the garden—stacking bales of hay, moving frames for raised garden beds, turning cover crops into the soil, hoeing weeds, or hand-pulling Bermuda grass.

You may just find that the community garden has become your new gym. Want to pump it up even more? Walk or ride your bike to and from the garden for extra exercise. Just be careful with that pitchfork.

MAKE YOUR LAWN LOOK GOOD ENOUGH TO EAT

The surge in backyard gardening in recent years is having an interesting ripple effect on front lawns as well. Some people are going all out with clearly visible vegetable gardens, but others are discovering the joys—and the bounty—of a less obvious option called edible landscaping. Edible landscaping involves incorporating edible plants, trees, and bushes that are also aesthetically pleasing into your landscape design. Done right, it tends to be more forgiving if you get a little lazy with maintenance, and it can include many perennial choices that will reduce your annual landscaping costs while also saving you money at the supermarket. Plus, you can add edibles a few at a time, so it's a strategy that fits into busy schedules and limited budgets. Here are some tips on how to convert your front yard into an edible landscape:

1. GO ONE FOR ONE. Replace an azalea bush with a rosemary or sage plant, both of which can grow quite large and provide you (and all your neighbors, and the food pantry) with fresh herbs throughout most of the year. Pull out the Carolina jasmine and put in a muscadine or grape vine near an arbor. Instead of planting your annual fall snapdragons, plant rainbow chard or red and speckled heirloom lettuces for rich fall color. Pull out the hedges and replace with a row of blueberry bushes. Put herbs around your mailbox instead of begonias and invite your neighbors to pick as they wish.

2. REDUCE THE LAWN. Little by little, year after year, choose a small part of your lawn and replace it with edibles. These can be easy-care perennial herbs that you add to expand your border beds a bit each year, or a few fruit and nut trees that turn a little-used corner of your property into your own private orchard.

3. LET IT SPREAD. Some herbs—such as mint, creeping thyme, and lemon balm—are perennial herbs that will take over if you let them. Well, why not let them? These perennials smell terrific when you brush past them or crush them under your feet, and there is really no limit to what you can make with them, from healthy home-made tea to mint chocolate chip ice cream.

4. HIDE IT. If you want the benefits of a raised-bed garden but don't want it to be so obvious on your front lawn, you can easily surround one or two beds with herbs or other edibles so that the frame is not even visible from the road.

★ TOP 10 ★

ROOTS AND TUBERS

Root crops are easy to grow and do best in deep, loose soil that has been well tilled. They nourish your brain and can be stored all winter long. Here are a few of my favorites:

1. **SWEET POTATOES:** You will find these tasty jewels on multiple Top Ten lists in this book because they just may be my favorite crop to grow—and eat. Plant the slips after last frost, eat the tender green tips, and harvest the tubers when mature. Sweet potatoes prefer a sandy soil but will tolerate heavy clay; they prefer full sun with lots of heat. For a tight space, try the Bush Porto Rico variety, or if you have room to let the vines roam, I recommend varieties like Beauregard or Georgia Jet.

2. **CARROTS:** These nutrient-rich roots are relatively easy to grow from seed, though they take a few weeks to germinate. Like all other taproots, carrots prefer to be sown directly in the garden. They will do best in a deep, sandy soil, preferably free of rocks. Carrots are a good crop to grow in raised beds and are good companions to lettuce, onions, and tomatoes. Some good varieties for home gardens include Nelson, Danvers, and Scarlet Nantes, and for mini carrots, try Thumbelina or Little Finger.·

3. **TURNIPS:** These underappreciated veggies are easy and fast to grow directly from seed and are delicious raw, roasted, or even pickled. Sow turnips in early spring or fall. My favorites are the small white varieties like Hakurei, Snowball, and White Egg.

4. **RADISHES:** If you are not very patient or need a quick confidence builder, radishes are your crop. These easy-to-grow veggies do best when direct sown in early spring or fall. They do not handle heat well, so make sure to get them planted early and sow in 1- to 2-week successions for an ongoing harvest. My favorite varieties for the home garden are French Breakfast, Easter Egg, and Red Meat.

5. **BEETS:** These health-giving vegetables are a little finicky to grow, and I have found they are one of the few root crops that do well grown indoors or in a cold frame, then transplanted outdoors right around last frost. Some of the best classic red beet varieties include Red Ace, Detroit Dark Red, and Merlin. For fancier specialty beets, try Chioggia, Cylindra, or Golden.

6. **POTATOES:** One of the easiest and most enjoyable crops to grow, potatoes are planted from eyes in early spring. They do best in light, sandy soil and can even be planted directly in straw. They like a moist but not too soggy soil. I like to grow a mix of fingerlings, Yukon Gold, Yellow Finn, and a classic red such as Red Pontiac or Red Norland.

7. **PARSNIPS:** Parsnips are an investment, as they take upwards of 120 days from seed to harvest, but they are well worth the wait and they get sweeter after a frost. Sow in early to mid-spring and keep moist until seeds germinate, which can take as long as 3 to 4 weeks. Parsnips are an excellent alternative to potatoes, as they are more nutritious with fewer calories. Some good varieties include Javelin, Harris Model, and Lancer.

8. **RUTABAGAS:** Rutabagas are easy to grow and store well through the winter. They like a well-drained, fertile soil, with a pH between 6.5 and 7. Plant in late summer and wait for a few frosts before harvesting to increase sweetness. A few of my favorite varieties are Helenor, Joan, and Laurentian.

9. **CELERIAC:** Celeriac is best sown indoors 12 to 14 weeks before last frost, potted into 2-inch pots, hardened off (transitioned to the outdoors using a cold frame), then transplanted into the garden. Harvest in late summer through fall, when the roots are between 3 and 5 inches in diameter. If treated correctly after harvest, the roots can last up to 8 months. My favorite varieties are Brilliant, Prinz, and Giant Prague.

10. **JERUSALEM ARTICHOKE:** Also known as sunchokes, these delicious perennials have a nutty, sweet flavor and can be eaten raw or cooked. They grow up to 8 feet tall, with beautiful daisylike blooms, and they make excellent windbreaks. Plant tubers in the fall and harvest the following year in late summer or late fall. Recommended varieties include Stampede and Challenger.

"The smallest seed of faith is better than the largest fruit of happiness."
Henry David Thoreau

4.

SOWING = FAITH

Few things are more rewarding than growing something from seed to harvest. It is astonishing to sow such a tiny speck in the earth and know that it will grow and reproduce a hundred more seeds, each doing the same in a perpetual cycle of abundant regeneration.

Planting a seed is both an act of independence and a leap of faith. Your crop is always, ultimately, at the mercy of Mother Nature. But there are techniques that will increase your odds of success and maximize your yield. In these pages you will learn the best way to plant seeds and seedlings, utilize space, dig holes, soak seeds, plant by the moon, and irrigate properly and wisely. In the process, you will become a true citizen farmer, as well as more successful with all the seeds you plant in life.

TOOLS FOR SOWING

SEEDERS: For seeding trays indoors or in a greenhouse, there are several types of hand seeders ranging in cost from $4 to $74. Most of your backyard seeding can be done by hand, but if you want to get a little more mechanized I recommend stepping up to the popular Earthway Vegetable Seeder, which comes with standard plates for different sizes of seed and can easily be adjusted to plant everything from carrots to corn. It is very quick and easy to operate, and at just over $100, it is a great investment. There are also some higher-quality precision single-row seeders that range from $150 to $200, as well as four- and six-row seeders for planting very tight beds of salad mix and other small- and medium-seeded crops. These are for the serious grower, as they are in the $250-to-$600 range.

HEAT MAT: If you are germinating warm-season veggies like tomatoes, peppers, and eggplant indoors or in a greenhouse, you will want to invest in a heat mat. Newly seeded trays need to be placed on the heat mat for only a few days until seeds sprout, then another rotation of trays can make use of the heat mat. This is valuable real estate, as each mat costs between $25 and $100. These mats will last many years and can be used over and over again, so they are a worthwhile investment.

SOIL BLOCKERS: These clever little devices allow you to make blocks of soil in a variety of sizes for seed starting. They take some getting used to and work best on wood trays that are often custom built, or you can use standard plastic trays made for this purpose. To use, moisten your germination mix and press the blocker into a container of the mix; release the blocks onto your tray as they are formed. You are then ready to plant the blocks with seeds, water, and grow. Prices for handheld blockers range from $30 to $100.

SEEDLING TRAYS: There are many types of trays in which to plant your seeds; hard plastic trays are the most popular among small-scale organic farmers. These come in all different sizes, from 50 to 200 cells per flat. (Cells are the little holes that hold one plant each.) You can also upcycle egg and milk cartons for this purpose, but these trays are very effective and worth the $5 to $10 investment. You should only need a few of them to get all of your backyard seedlings going, so I say buy the best quality and use them for years. It is recommended to sterilize them with hot water between uses.

TRANSPLANT POTS: Once your seedlings outgrow their starter trays, they are ready to either go directly into the garden or be transplanted up to a bigger pot. You can find 2- to 4-inch pots made out of everything from coir, peat, and recycled plastic to cow manure. They all have their pros and cons, but I recommend going with either recycled plastic or peat. (The peat is nice because you do not have to take the plant out of the pot before transplanting, which reduces transplant shock, and there's no pot to clean or throw out. The downside is that peat is not a totally sustainable resource.) The most important thing here is healthy plants, so don't worry too much about the type of pot; focus more on soil, watering, and removing the plant from the pot before it gets root-bound.

SOIL THERMOMETER: A great way to tell when it's time to plant your veggies in the garden is by taking the soil's temperature. Seed packets often list the ideal soil temperature for germination, so a soil thermometer allows you to determine what can be planted in the garden when. Another good use for a soil thermometer is to test the temperature of the soil in your seedling trays and adjust the heat mat accordingly. The cost of knowing your soil temperature is a whopping $12. Well worth it.

WATERING CAN: This may seem obvious, but not all watering cans are created equal—in fact, some can do more damage than good. For seed starting, you want a watering can that has a fine spray and does not leak droplets out of the spout; even one droplet falling onto a newly planted tray can wash out a few seeds. I highly recommend investing in a high-quality watering can from brands like Dramm or Haws. Be prepared to spend about $30.

FOGG-IT NOZZLES: These well-made nozzles connect to a hose and create a dewlike mist for watering tender seedlings. They are great for providing a gentle supply of water as well as knowing how much water you are actually applying. Fogg-It offers nozzles in superfine, ½-gallon per minute (gpm), 2 gpm, or 4 gpm, all for just $8.99 each.

LIGHTS: If you do not have a greenhouse or a bright window in the house large enough for all your seedlings, then some lights would be a good idea. The technology for indoor grow-lights has been improving rapidly, and today you can set up a T5 2-foot system for about $60 and a 4-foot system for about $100.

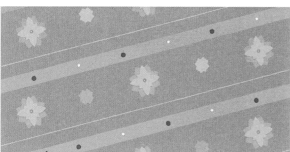

TIPS FOR PLANTING SEEDS IN TRAYS

CHOOSING SEEDS: Looking through rich, colorful descriptions of the vegetables, flowers, fruits, and herbs you plan to bring to life in your garden can be incredibly inspirational—and daunting. Choose wisely and order the appropriate quantities for your scale, keeping in mind successional plantings and space limitations. Seeds are relatively cheap, and I like to have plenty on hand. You will learn to save seeds over time and therefore save money, develop strains better acclimated to your garden, and strive toward self-reliance.

You have numerous choices for seeds, all of which can be easily accessed online or bought at many local garden centers. (See the resources section for a list of some of my favorite seed companies.) Hybrid seeds are varieties that have been developed to be resistant to diseases and to produce certain desired characteristics. Some may be treated or genetically modified, so pay attention to labeling to avoid these. Heirloom seeds are age-old favorites passed down from generation to generation. Both hybrid and heirloom seeds can be either organic or not organic, depending on how the host plant was grown.

Determinate tomato plants will grow to a pre-determined size with fruit that matures around the same time. Indeterminate varieties grow and ripen continually; they tend to need more room and staking. Determinate types are particularly useful if you are growing tomatoes to make sauce, as you will have a greater bounty at one time. Indeterminate types are better for maintaining a supply of salad or sandwich tomatoes throughout the season. Most gardeners tend to plant some of each.

SEED SOAK: Just before planting, put the seeds in a bath of water with a little kelp, seaweed, and/or biodynamic preps mixed in. When they have soaked for a few hours, pour the seeds out onto a paper towel and let them dry just enough that they are easy to handle. It is best to sow seeds while they are still moist.

SOIL: Mix your soil ingredients well in a clean bucket for home gardeners, or a large container such as a tub or wheelbarrow for bigger gardens. Add water to the germination mix using a watering can with a rain spout, or use a hose with a sprayer wand. You do not want to pour water in, but rather give a rainlike sprinkling as you further mix the soil to avoid creating clumps. Moisten the soil slowly, in spurts, to the point that when you squeeze a handful tightly, just a drop comes out. You can always buy a premixed germination soil if you want to skip the mixing step; just make sure the moisture content is about one drop to a tight squeeze.

TRAYS: Once you have the soil well mixed and at the right moisture level, it is time to fill the seed trays. I like to use my hands to scoop the soil and fill the trays. Gently pack the soil to make sure there are no air pockets in the bottoms of the cells. Most trays fit into each other, so you can place another tray on top of your filled one to push down the soil evenly throughout. This will also make little indentations in each cell from the drainage hole in the bottom of the pressing tray.

SOWING: Fill your hand with a bunch of seeds and begin to place one or two seeds in each cell. The general rule of thumb is to plant the seed as deep as it is big. Keep in mind some seeds require light to germinate, so be sure to read the instructions on the seed packet. The germination rate listed on the packet, which will appear as a percentage, lets you know approximately how many seeds to put in each cell. The higher the percentage, the fewer seeds go in each cell. Anything 90 percent or higher means one seed per cell. If you overplant now, it will take more time later to thin, and that pulling on the roots can harm the seedlings. Multiple seedlings also have to compete for water, nutrients, and sunlight.

I like to fill a bunch of trays with soil before starting to plant seeds, so that I am able to perform the two processes without interruption. This also helps keep your hands clean when planting seeds; going back and forth from filling trays to planting seeds is not ideal. Do not fill more trays than you sow, because they will dry out if left too long.

Plant each seed firmly in the soil as close to the middle of the cell as possible. Some people like to make small indentations in the center of each cell, either with a finger or pencil or by using the tray-on-tray technique.

After all the seeds are planted in each tray, broadcast a light covering of vermiculite over the entire tray; this is an easy and effective way to cover your newly planted seeds. Another option is a light covering with more germination mix and a very gentle patting down so the seeds make good contact with the soil. For the first few days, the trays can be in darkness, but once the seeds sprout they need direct access to sunlight so they don't get leggy as they stretch in search of light. Make sure to keep the sprouts moist until they have a few true leaves and enough roots to withstand a short dry spell between waterings.

WATERING: The last step after sowing your seed trays is watering. Gently water using a watering can with a sprinkler spout or a hose with a spray nozzle or mister nozzle. Be very careful not to water too hard, as this will move the seeds to the edge of the cell. Once watered, place the trays in your window, greenhouse, or germination box, making sure to water consistently so as not to let the soil dry out.

DAMPING OFF: One of the biggest problems in starting seeds is something called damping off, which is when the area where the stem meets the soil rots and the seedling dies. This often results from watering too lightly, causing the water to sit on top of the soil rather than soaking into the depths where roots will grow. Avoid this by watering properly. Another tactic I have learned for reducing the risk of damping off is to add oak leaves and eggshells to the bottom of each tray, as they provide the roots with extra calcium and fungi. This is easier done in an open tray than in a tray with cells and also works well with soil blocks.

TIPS FOR PLANTING SEEDS AND SEEDLINGS DIRECTLY IN THE GARDEN

Let's imagine you're finally starting a garden. You've built raised beds in a nice, sunny spot, you've balanced your soil, and now you are ready to plant seeds. Here are a few basic tips for growing from seeds or seedlings:

1. PICK THE RIGHT TIME. The ideal time for planting depends on your crop choice and your specific climate. Most seed packages recommend when to plant specific seeds relative to your area's first frost date and suggest ideal soil temperatures for germination as well.

2. CLEAR AND PREPARE A SPACE. Before you start a new project at work, it's a good idea to take time to file away old projects and organize your work space. The same principle applies when you plant seeds; you need to make room for them. If they are fall seeds, that means removing summer plants that are done for the season, loosening up the soil a bit, and mixing in some compost and organic fertilizer before planting.

3. PLANT SEEDS AS DEEP AS THEY ARE BIG. For instance, a garlic clove would need to be planted about 2 inches deep; a zucchini seed, about half an inch deep. Tiny lettuce seeds could simply be tossed on top of the soil and lightly dusted with a soil covering or gently raked into the top ¼ inch of the seed bed. Hilling (which means mounding soil in little hills) is recommended for melons, cucumbers, and squash. Planting three seeds of the same variety in each hill gives you the opportunity to choose the strongest of the three to continue to let grow. You'll also hear the term "hilling up" in relation to potato plants. As potato plants grow, you add soil around the stem so that a greater number of roots will form, from which more potatoes will grow. You can use wheat straw to hill up instead, which makes for clean, easy harvesting, but be sure the potatoes don't get exposed to the sun or they will develop poisonous green "shoulders."

4. KEEP MOIST. Be sure to give your "babies" the TLC they need as they grow. Seeds need to stay moist until they germinate, which usually takes 7 to 10 days, so plan on watering them daily. You can do this by hand with a watering can or hose nozzle, or set up an irrigation system. This is a good time to use a sprinkler on a timer, especially if you travel or have an erratic schedule.

5. IN GENERAL, DON'T WORRY. Seeds want to grow, and as long as conditions are right for them, they will. If they end up too close, you can move some plants or cut some of them off at the soil line so the remaining ones have room to grow (both of those methods are called "thinning"). If you're really lucky, some of your favorite plants from this year will have dropped some seeds to grow next year. These little gifts of the garden are called "volunteers." You can usually move them around if they are not in the place where you want them, or you can let nature take its course.

6. TRANSPLANT WITH CARE. When transplanting seedlings, be very gentle and transplant at a mild time of the day, such as early morning or late afternoon into the evening. A cool, cloudy day with a little mist or rain is ideal. A full moon, especially if it's in a root sign, is also an ideal time to transplant seedlings, as the roots will grow quickly, helping the plant to get established with minimum transplant shock.

PLANTING WITH THE MOON

The gravitational pull of the moon affects the tides. It makes sense, therefore, that the moon's relationship to the earth most likely affects groundwater as well. This is a basic principle of biodynamic gardening and farming, and it's also an age-old strategy that our ancestors learned to use after witnessing the results firsthand. Why not try it in your garden and see if it makes a difference? Here are the basics:

WHEN THE MOON IS WAXING: As the moon starts to grow bigger in appearance in the night sky, its gravitational pull is stronger. This is a good time to encourage growth in your plants. Recommended actions are transplanting, grafting, fertilizing, and harvesting leafy greens and fruits that you intend to consume right away, as their water content is going to be higher during this time.

WHEN THE MOON IS FULL: That big, bright full moon is your sign that it's the ideal time to plant seeds or transplant seedlings. The high moisture level in the soil will aid in germination, and the moon's strong influence will help roots and leaves to grow quickly.

WHEN THE MOON IS WANING: As the moon diminishes in appearance in the night sky, the moisture level in the soil recedes. Sound like a bad thing? Not at all! This time of the month is the perfect time for planting seeds that do most of their growing underground, such as root crops, or other plants that depend on robust root growth, such as trees. This is when you want to prune plants and divide perennials. Most important, after all your hours of hard work tending the garden, this is when you want to harvest any fruits or vegetables that you intend to dry or root crops you intend to store, as their water content will be lower than during the waxing or full phases of the moon.

WHEN THERE IS A NEW MOON: When no moon is visible in the night sky, the water content of your soil is lowest, so pay extra attention to watering your garden, as your plants will rely on you at this time. This is also a good time to plant seeds.

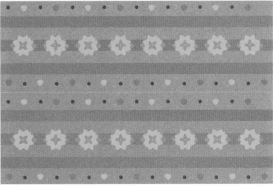

THE CONSTELLATIONS AND FOUR ELEMENTS: The moon also has a strong influence on the earth through the twelve constellations. There are four elements, each with a direct relationship to the twelve constellations. When the moon is in one of these signs, it is a good time to plant, cultivate, or harvest the associated crop type:

- **FIRE SIGNS (*ARIES, LEO,* AND *SAGITTARIUS*)** — for fruiting crops
- **EARTH SIGNS (*TAURUS, VIRGO,* AND *CAPRICORN*)** — for rooting crops
- **AIR SIGNS (*GEMINI, LIBRA,* AND *AQUARIUS*)** — for flowering crops
- **WATER SIGNS (*CANCER, SCORPIO,* AND *PISCES*)** — for leafy crops

WHAT TO PLANT WHEN YOU HAVE FAITH—BUT WANT IT FAST

It is said that planting a seed is the ultimate act of faith—especially in certain cases. Some seeds take their sweet time, including garlic, onions, and brassicas such as cabbages, broccoli, and Brussels sprouts. You could birth an elephant before leeks are ready—I've had some that have taken two years! For those of us used to a fast-paced, instant-gratification lifestyle, this waiting game may seem unbearable. Here are some ways to cope:

1. PLANT A WIDE VARIETY OF FRUITS AND VEGETABLES. If you are busy harvesting daily salads, you won't even notice how long the carrots are taking. Since your radishes take just 28 days from seed to plate, you may not mind that the beets or turnips take a bit longer. If you're cutting collard leaves, you won't mind waiting for rutabaga.

2. FOCUS ON FAST-GROWING CROPS. Still waiting for those tomatoes to turn red or the watermelons to ripen this summer? Or maybe you're gardening with a child whose patience is stretched even waiting for popcorn to pop? You might want to focus solely on fast-maturing crops such as radishes, leafy greens, and summer squash for a while. To get an even bigger jump, go straight to transplants, and you may be able to snip some lettuce leaves or sneak a strawberry or two pretty quickly.

3. GO WITH PERENNIALS. This may sound counterintuitive, but stay with me a moment: Why not plant some fruits and vegetables that take a long time to be ready to harvest, like fruit trees and asparagus, but which, once established, just require minor care and pruning, and then deliver like Old Faithful? The trick is choosing varieties that ripen at different times so you have a staggered harvest.

4. ALLOW YOURSELF TO GROW IN FAITH. Sometimes the best way to deal with the waiting game in gardening is to give in to it and allow your faith to expand. You may even find that anticipation becomes your favorite part of the journey, and your belief in a bountiful outcome may spread to other parts of your life as well.

WAYS TO SOW SEEDS OF CHANGE IN YOUR COMMUNITY

What greater metaphor for change is there than planting a seed? You water it, and before long, it grows. Over time, it bears fruit. Finally, the wind and birds carry its seeds near and far, and more and more grows. Change in our communities is like that, too.

Consider this little story as an illustration. I drive by a certain home frequently, and a few years ago, a row of tomatoes suddenly appeared on the front lawn. Over time, I noticed the row got longer, and peppers had been added. Last year, I saw greens growing in the spring and fall. Just yesterday, I noticed that the tomatoes had been removed for the season and fall crops were being planted. But then something amazing happened. As I was smiling to myself about that garden's success over the years, I noticed the next-door neighbor's home. Where previously there had been only lawn, there was now a freshly tilled row, ready for planting. The first gardener's idea had clearly grown.

The moral of the story? Every little seed you plant matters, in ways you may not know. Here are some ways you can sow seeds of change in your community:

1. PLANT TO SHARE. If your neighborhood has a common area, ask to be in charge of it, or at least get permission to plant a few things. Choose edibles like herbs and let your neighbors know they are there for the picking. Oregano, rosemary, lavender, lemon thyme, mint, lemon balm, and chives are all low-maintenance perennials that do well in many climates and are pickable during a good portion of the year. Of course, this idea will also work well right around your mailbox.

2. GARDEN WHEN PEOPLE CAN SEE YOU. Do your front-yard gardening when you know people will be passing by, such as early-morning or after-work joggers or dog-walkers. This gives you the opportunity to wave to them, or, if you have time, to chat with them, answer questions, share clippings, and encourage their interest in gardening. Consider doing your gardening especially when school buses go by, as little eyes are watching and learning from you more than you may realize.

3. SHOW UP AT CITY HALL. Speak up in support of the amazing urban agriculture initiatives going on in your city and around the nation. Write blog posts, add photos of gardens you visit on social media, and attend meetings where your voice may make a difference.

4. GET OUT IN THE COMMUNITY. Sign up for a crop mob, where you go out to a local farm and help knock out a bunch of chores with a group of other volunteers. Donate some of your time to a nonprofit organization that is cultivating gardens in low-income schools, churches, or boys' and girls' clubs.

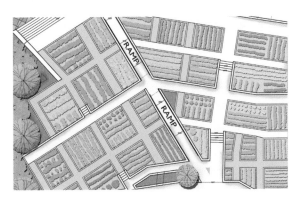

★ TOP 10 ★

EASIEST CROPS TO GROW FROM SEED DIRECTLY IN THE GROUND

Seeds have one of the best returns on investment out there. Buy a $3 packet of squash or cucumber seeds and watch your investment grow exponentially. Here are some of the easiest crops to seed directly in the ground:

1. **RADISHES:** Plant seeds ½ inch deep in 2- to 3-inch bands about 1 inch apart in rows 6 to 12 inches apart. Plant in early spring and fall, as radishes do not like heat. Keep moist and pick as soon as they are ready, as they will quickly go from tender, juicy, and sweet to pithy, tough, and spicy.

2. **SUMMER SQUASH:** After the danger of frost has passed and soil temperatures reach 62°F to 70°F, plant approximately 3 seeds per foot about ½ inch deep in rows 4 to 6 feet apart. Thin to 1 plant per foot when seedlings are 4 to 6 inches tall. It is possible to transplant these thinned plants, though squash is very sensitive to transplanting, so make sure to take extra care.

3. **CUCUMBERS:** Wait until soil temperatures reach 70°F, then sow 3 seeds per foot about ½ inch deep in rows 4 to 6 feet apart. Thin to 1 plant per 8 inches and cover with floating row cover (see page 136) as soon as you plant to protect against cucumber beetles.

4. **BEANS:** Many kinds of beans can be direct sown after frost, when soil temperatures reach between 70°F and 90°F. Sow seeds about 1 inch deep and 2 inches apart in rows 18 to 36 inches apart. Plant every 2 weeks during summer months for a continual harvest. For optimal performance, inoculate seeds immediately before planting.

5. **PEAS:** A cool-season crop that can be planted in early spring as soon as the soil can be worked. Sow seeds ½ to 1 inch deep and 1 to 2 inches apart in a 3-inch band. For dwarf varieties, plant in rows 12 to 18 inches apart and do not trellis. For taller varieties, plant in rows 3 to 6 feet apart and trellis soon after planting. For optimal performance, inoculate seeds immediately before planting. Try again in early fall for a late-season harvest.

6. **SUNFLOWERS:** Plant after last frost, when soil temperatures reach about 70°F. Sow seeds ½ to 1 inch deep and 8 to 24 inches apart in rows a few feet apart. There are many different sizes of sunflower, so refer to the seed packet for more specific spacing recommendations.

7. **CILANTRO:** A cool-season herb best planted every few weeks starting in early spring, just after last frost. The summer heat will cause bolting, so continue planting again in fall when temperatures start to cool down again. Sow seeds about ½ inch deep and 2 to 4 inches apart in rows 12 inches apart.

8. **ARUGULA:** A cool-season leafy green that can be direct sown in early spring after the last frost and planted every 2 to 3 weeks until it's too hot. Continue to sow again in fall when temperatures cool; in mild climates it can overwinter (continue to grow through the winter). Sow about 20–40 seeds per foot ⅛ inch deep in rows 8 to 12 inches apart.

9. **LETTUCE:** A cool-season favorite that can be direct sown after last frost through summer and again in fall. Sow ⅛ inch deep, as lettuce does require a minimum amount of light to germinate. For baby lettuce, plant about 60 seeds per foot in 2-inch bands with about 1 inch between bands. For head lettuce, plant about 3 seeds every 8 to 10 inches with 12 to 18 inches between rows.

10. **CARROTS:** Plant in early spring or late fall, when soil temperatures are between 55°F and 75°F. Sow 30 seeds per foot ¼ to ½ inch deep and thin to 1 to 2 inches in rows 18 to 24 inches apart.

"Be not afraid of growing slowly, be afraid only of standing still."
Chinese proverb

5.
GROWING = PATIENCE

It can be a long wait from seed to harvest. The corn. The tomatoes. The eggplants. Let's not even talk about pumpkins.

Summer gardening is a far cry from spring gardening, when having fresh lettuce leaves for that evening's dinner can be as easy as walking out your door and snipping the outer leaves so the rest of the plant can keep growing for the next day, and the next, and the next.

Hang in there. If past experiences from seasoned gardeners are any indication, by the time you are ready to hit the pool, you will be knee-deep in cukes and zukes, whether you grow them or not. A friend or neighbor is sure to be leaving them at your doorstep, or in the open window of your car if you're not careful.

In the meantime, there are valuable lessons in delayed gratification that can be learned by watching and participating in the process of growing a garden from an early age. In our fast-paced, technology-driven modern-day reality, it's more important than ever for us to slow down and connect to the natural rhythms of a plant's growth. It teaches us to be more mindful and patient stewards of everything we want to cultivate in life.

This chapter will cover all aspects of growing a healthy garden once your seeds are sprouted, including fertilizing, weeding, mulching, pruning, trellising, and more. I will also show you how you can apply these techniques to nurture other areas of your life.

★

TOOLS FOR GROWING

HAT: Working out in the sun for long periods of time can be harmful to your skin. I recommend choosing a hat that breathes and has a large enough brim to give you some relief from the sun.

HORI HORI WEEDER AND ROOT CUTTER: This is a popular Japanese tool for transplanting, weeding, and cutting roots. It is many gardeners' favorite tool because of its versatility and durability. It costs about $30.

HOES: There are many different kinds of hoes for specific chopping and slicing tasks. The inexpensive hoes you find at a hardware store are likely to break within the first year of use. A good hoe can save your back and actually make weeding enjoyable. I love the Rogue hoes, Glaser stirrup hoes, and DeWit diamond and half-moon hoes. There is also a line of tools from a German company called WOLF-Garten in which you can use one handle with a variety of different interlocking attachments; the draw hoe is my favorite. These quality hoes range from $30 to $100.

WHEEL HOE: If you find you have more rows than you can weed with a stirrup hoe, it may be time to step up to a wheel hoe. This is essentially a stirrup hoe with a long handle and front wheel so you can fly down rows without the back-and-forth motion required with a long-handled hoe. They are a big investment and only justified if you have some serious weeding to do. Prices for decent wheel hoes range from $250 to $400.

PRUNING SHEARS: A good pair of pruners is essential. You can go cheap and buy a $15 pair of pruners that will last a few months, or go high-quality with a Felco #2 and you'll never have to buy another pruner again. They retail for about $60 and a replacement blade costs $15.

GLOVES: Whether it's because you don't want to get dirt under your nails or you're tackling a thorny raspberry patch, a pair of gloves will serve you well. Many tasks are done faster and more easily when we don't have to worry about cold or blistered hands. I recommend having a pair of leather gloves, which shouldn't cost more than $10 or $20. If you grow roses or other thorny crops, a pair of professional rose gloves, at about $30, is a must.

KNEELING PAD: It is inevitable that you will end up on your knees in the garden, and not just to pray for rain. In organic gardening, weeds are often pulled by hand, and to avoid hurting your back, resting on your knees is often the best option. Some people like to take kneepads with them wherever they go by wearing them, and others prefer a pad that they move around as needed. Both options range from $10 to $35.

SPRAYER: For applying liquid fertilizers, foliar sprays, and compost tea, you will need a decent sprayer. For very small-scale gardens, you can get away with a pump sprayer for $25 to $50. I prefer a backpack sprayer, as it keeps both your hands free and the weight is on your back rather than in your hand. A good-quality 4-gallon Solo or similar backpack sprayer will cost about $100.

COMPOST-TEA MAKER: Applying compost tea to your soil and plants is a great way to make your garden thrive. Compost tea is very rarely found at retail garden stores; most people make their own and there are a variety of methods, from a simple 5-gallon bucket and fish-tank pump to high-tech commercial options. A homemade version using your own compost can cost as little as $30, a kit can be bought online for between $60 and $200, and a small-scale commercial tea brewer costs about $600. (See Compost Tea recipe, page 47.)

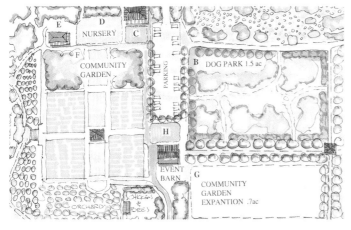

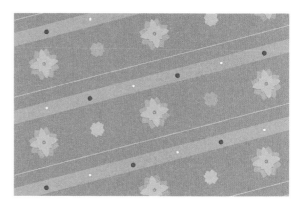

"A good gardener looks at every plant every day."
Alan Chadwick

GROWING SEASON TIPS

The best thing a gardener can do is be present with his or her garden, watching it closely on a regular basis. Here are some quick tips to help you keep up with your garden:

DAILY OBSERVATION: Take a walk around your garden and check for signs of disease, decay, and destruction. Remove dying leaves. Squash the eggs of harmful insects (use the Internet or a book like *Rodale's Garden Insect, Disease and Weed Identification Guide* to distinguish the bad eggs from the good) and handpick bugs from leaves (drop them in soapy water). This means the dreaded tomato hornworm as well. Don't want to touch the squishy fellows? Snip the branch on which they are perched right into a cup of soapy water or feed them to your chickens.

MULCHING: Covering the soil in the garden helps preserve moisture, suppress weeds, and build organic matter around crops. Possible drawbacks are creating a habitat for bugs and decomposition near the bases of plant stems. Mulch can also be messy if applied on top of a crop like lettuce. Its benefits, however, far outweigh its hazards. A mix of grass-hay mulch and aged bark or wood helps provide a "warm blanket" of beneficial bacteria and fungi.

TRELLISING: Providing plants with the necessary support is an important step in staving off disease and rodents, and it dramatically increases your chances for success. Add trellises, bamboo poles, and even an old folding ladder or two to your garden to provide reinforcement and shaping. Also, spend a few minutes helping your plants know where they should go. Direct vines to fences, poles, trellises, and other supports you may have already added to the garden by wrapping their little tendrils around the support. A well-managed bamboo clump can provide all the trellises you need, but make sure to pick varieties that clump rather than spread.

PICKING: Most plants like to be picked often and will actually increase production when you do this. Once crops like beans, cherry tomatoes, and lemon cucumbers start producing, they will be happy to make it into school and camp lunch boxes daily. Check your fruit trees, too, as you might be surprised to find ripe fruit hiding beneath the leaves. Pick okra often and at your own risk—those plants produce like nothing you've ever seen! If you find yourself with surfeit produce, you have options: pick and share them with your local food pantry; pick and pickle or dehydrate them to save for your winter menu; or let them dry and save them for their seeds or for holiday crafts.

PRUNING: While pruning is primarily used for fruit trees and berries, there are a few veggies that appreciate a clip here and there as well. Pruning tomato suckers helps maximize fruit production and reduce risk of disease and pests. Suckers (new small stems that begin to grow where a branch meets the main stem) drain energy from the plant, making for lots of small tomatoes rather than a good yield of nice, full ones. Other crops that like to be pruned include roses and other flowers that need deadheading, as well as perennial plants that get cut back at the end of each season.

BENEFITS OF CROP ROTATION

One of the most effective ways to deal with pests and disease organically is by rotating crops. This practice can be a little difficult in a backyard garden due to limited space, but it should be part of your ongoing maintenance plan. Here are some key benefits of rotating crops:

PEST AND DISEASE CONTROL: Plants from the same families tend to be attacked by similar pests and diseases. Moving crops around the garden according to plant family can help decrease the population of pests and buildup of certain pathogens in the soil.

REDUCING SOIL EROSION: A common mistake is to leave soil exposed for long periods of time. A benefit of crop rotation is that planting crops back to back prevents soil from being dried up by the sun or washed away by rain. This is especially true with larger farms that leave fields fallow for long periods of time, which can result in serious runoff and nutrient leaching. Cover crops help hold the soil together thus preventing any loss of topsoil due to wind, water, or sun. They also improve the overall fertility, biology, and structure of the soil.

NUTRIENT MANAGEMENT: Some crops take more nutrients out of the soil than others, with the hungriest of them being referred to as "heavy feeders." "Heavy givers" include cover crops and green manures, which help put nutrients and organic matter back into the soil between crop cycles. A good rotation for nutrient cycling is to plant a heavy giver, followed by a light feeder, followed by a heavy feeder, then repeat the cycle. See the list on page 113 for examples of heavy, medium, and light feeders.

CROP ROTATION PLANS

There are many different ways to rotate your crops, and while an eight-year rotation is ideal, this is not realistic on a small scale so I have focused here on four-year models, with some basic seasonal tasks for each section.

BASIC FOUR-YEAR ROTATION

	SECTION ONE	SECTION TWO	SECTION THREE	SECTION FOUR
YEAR ONE	Legumes	Brassicas	Nightshades	Onions and root crops
YEAR TWO	Brassicas	Nightshades	Onions and root crops	Legumes
YEAR THREE	Nightshades	Onions and root crops	Legumes	Brassicas
YEAR FOUR	Onions and root crops	Legumes	Brassicas	Nightshades

ADDITIONAL NOTES:

- Incorporate cucurbits (cucumbers, melons, and squash), lettuce, and corn wherever convenient; avoid growing them in the same place too often.
- Perennial crops such as asparagus and rhubarb do not fit into the rotation.
- For a three-year rotation, combine legumes with onions and roots in one section, follow with brassicas, then nightshades.

ALTERNATIVE FOUR-YEAR ROTATION

	SECTION ONE	SECTION TWO	SECTION THREE	SECTION FOUR
YEAR ONE	Brassicas and legumes	Nightshades	Root crops	Lettuce, spinach, greens, squash, and corn
YEAR TWO	Lettuce, spinach, greens, squash, and corn	Brassicas and legumes	Nightshades	Root crops
YEAR THREE	Root crops	Lettuce, spinach, greens, squash, and corn	Brassicas and legumes	Nightshades
YEAR FOUR	Nightshades	Root crops	Lettuce, spinach, greens, squash, and corn	Brassicas and legumes

SEASONAL TASKS

	SECTION ONE	SECTION TWO	SECTION THREE	SECTION FOUR
SPRING	Plow under cover crop	Add compost	Plow under cover crop	Add compost and organic granular fertilizer
FALL	Mulch	Amend soil based on tests and plant cover crops	Mulch	Plant winter cover crop

CROPS BY SEASON

- **MOST COMMON WARM-SEASON CROPS:** Beans, cucumbers, eggplants, melons, peppers, pumpkins, squash, corn, sweet potatoes, tomatoes

- **MOST COMMON COOL-SEASON CROPS:** Beets, broccoli, Brussels sprouts, cabbage, carrots, cauliflower, celery, greens (tatsoi, bok choy, mizuna), kale, lettuce, onions, parsley, peas, potatoes, radishes, spinach

CROPS BY FAMILY/GROUP

LEGUMES			
• Peas	• Beans	• Lentils	• Cover crops like clover

BRASSICAS			
• Cabbage	• Radishes	• Collards	• Brussels sprouts
• Kale	• Turnips	• Broccoli	

ONIONS AND ROOT CROPS			
• Onions	• Beets	• Carrots	• Parsnips
• Garlic	• Leeks	• Shallots	

NIGHTSHADES			
• Tomatoes	• Peppers	• Eggplants	• Potatoes

CROPS BY NUTRIENT NEEDS

The following list will help you better understand how to rotate crops based on fertilizer needs or nutrient richness:

HEAVY FEEDERS

These nutrient-hungry veggies require approximately 3 pounds each of nitrogen, phosphorus, and potassium per 1,000 square feet:

- Asparagus
- Beets
- Brassicas (broccoli, Brussels sprouts, cabbage, cauliflower, collards, kale, kohlrabi, radishes)
- Celery
- Corn
- Cucumber
- Lettuce
- Melons
- Onions
- Parsley
- Potatoes
- Pumpkins
- Rhubarb
- Spinach
- Squash
- Sunflowers
- Tomatoes

MEDIUM FEEDERS

These herbs and veggies require approximately 2 pounds each of nitrogen, phosphorus, and potassium per 1,000 square feet:

- Beans
- Chard
- Eggplants
- Greens
- Herbs (annuals)
- Okra
- English peas
- Peppers
- Root crops (carrots, garlic, leeks, parsnips, rutabaga, shallots, turnips, sweet potatoes)
- Strawberries

LIGHT FEEDERS

These crops are tolerant of less fertile soils and require only about 1 pound each of nitrogen, phosphorus, and potassium per 1,000 square feet:

- Bulbs
- Herbs (perennial)
- Southern peas

HEAVY GIVERS / SOIL BUILDERS

- Alfalfa
- Beans
- Clover
- Peas
- Rye, oats, and wheat
- Sorghum-sudangrass
- Sunhemp
- Vetch

COMPANION PLANTING

When planning your garden, it is a good idea to think about putting plants together that are going to complement rather than compete with one another. There are crops that grow well next to each other in the garden, and others that can actually harm each other.

FRIENDS AND FOES

Here is a list of some of the most popular veggies and the crops they like and don't like. Keep these good and bad combinations in mind both when you are planning and when you are rotating crops in the garden.

BROCCOLI

FRIENDS: alliums, beets, borage, buckwheat, calendula, carrots, chamomile, dill, mint, nasturtiums, and rosemary

FOES: mustard, peppers, strawberries, and tomatoes

BEANS

FRIENDS: beets, brassicas, carrots, corn, cucumbers, eggplant, peas, potatoes, radishes, squash, strawberries, and tomatoes

FOES: alliums, fennel, peppers, and sunflowers

CARROTS

FRIENDS: beans, leeks, lettuce, onions, peas, potatoes, radishes, rosemary, sage, and tomatoes

FOES: dill and parsley

CABBAGE

FRIENDS: beans, celery, cucumbers, dill, kale, lettuce, onions, potatoes, sage, spinach, and thyme

FOES: strawberries and tomatoes

EGGPLANT

FRIENDS: beans, lettuce, peppers, potatoes, spinach, and tomatoes

FOES: none

CUCUMBERS

FRIENDS: beans, cabbage, cauliflower, corn, lettuce, peas, radishes, and sunflower

FOES: melons and potatoes

KALE

FRIENDS: buckwheat, cabbage, and nasturtiums

FOES: pole beans and strawberries

GARLIC

FRIENDS: fruit trees, roses, and tomatoes

FOES: beans, cabbages, peas, and strawberries

LETTUCE	
FRIENDS: beans, carrots, radishes, strawberries, and kohlrabi	**FOES:** cabbage, celery, and parsley

ONIONS	
FRIENDS: beets, broccoli, cabbage, carrots, lettuce, parsnips, peppers, potatoes, spinach, and tomatoes	**FOES:** beans and peas

PEAS	
FRIENDS: beans, carrots, corn, cucumbers, potatoes, radishes, and turnips	**FOES:** alliums

PEPPERS	
FRIENDS: basil, carrots, eggplant, onions, parsley, spinach, and tomatoes	**FOES:** beans, fennel, and kohlrabi

POTATOES	
FRIENDS: basil, beans, beets, brassicas, corn, eggplant, horseradish, nasturtiums, peas, and squash	**FOES:** apples, celery, cherries, cucumbers, pumpkins, raspberries, sunflowers, and tomatoes

RADISHES	
FRIENDS: basil, carrots, cucumbers, eggplant, lettuce, melons, onions, nasturtiums, peas, root crops, spinach, squash, and tomatoes	**FOES:** beans and kohlrabi

TOMATOES	
FRIENDS: asparagus, basil, beans, borage, carrots, celery, dill, lettuce, melons, nasturtiums, onions, parsley, peppers, radishes, spinach, and thyme	**FOES:** beets, brassicas, corn, kale, peas, and potatoes

ZUCCHINI AND SUMMER SQUASH	
FRIENDS: corn, beans, nasturtiums, marigolds, onions, radishes, melons, and mint	**FOES:** Irish potatoes

MODEL CITIZEN PLANTS

Here are a few plants that are great to have scattered all around the garden, as they get along with everyone!

- Basil
- Marigolds
- Marjoram
- Oregano
- Thyme
- Tarragon

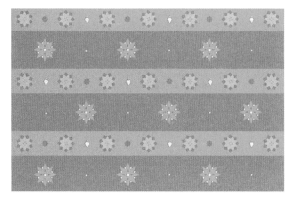

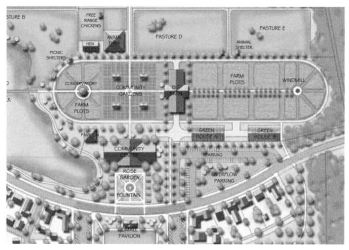

FOLIAR FEEDING: A HEALTHY WAY TO SPRAY YOUR PLANTS

There's a little secret about growing a healthy garden that you may not know yet, but it has the potential to enliven your plants almost immediately. It's called foliar feeding, a process of applying various organic fertilizers in liquid form directly to leaves via a spray bottle or a backpack sprayer.

I know this sounds counterintuitive, especially if you've been conscientiously watering just the roots and not the leaves (especially of tomato plants) so that you don't encourage disease. However, foliar sprays can alleviate magnesium deficiencies, deliver other trace minerals and natural hormones, and bathe a plant in beneficial bacteria and fungi that help fight off disease. The leaves absorb the sprays quickly through their stoma, openings that let in carbon dioxide and let out water and oxygen. Studies show that nutrients delivered via foliar feeding are evident in all parts of the plant within mere hours, and the positive impact of these nutrients is visible on the plants within less than a week. That means sickly plants might be bright green and healthy looking again before you know it.

HOW TO CRITTER-PROOF YOUR GARDEN

Those of us who garden have many tales to tell about rabbits, birds, deer, gophers, chipmunks, squirrels, and any number of other critters who snag our juiciest tomato or ripest fig, or dig up seeds before they even get a chance to germinate. Although I try to take a holistic view of gardening and accept that I am just a tiny part of a larger ecosystem, I have to admit I want at least some of what I plant to end up on my dinner table.

There is a long list of resourceful deterrents that farmers and gardeners have tried—with varying success—to save their harvests from a wide range of critters. These include human hair, fox urine, shiny compact discs dangling from stakes, traps, sprays, fake owls, bars of soap, and, of course, scarecrows. Here are some tried-and-true ways to foil the feathered and furry:

1. IF RABBITS ARE EATING YOUR GARDEN: Plant clover. This is the number-one tip I've heard for keeping rabbits away from your other delectables. They simply love clover and will choose to eat it over just about everything else. Toss some clover seeds between all your beds to give your backyard wildlife something to nosh on instead of your fall greens and carrots. Another strategy is to use raised beds and surround the area with a chicken wire fence, remembering to bury the wire a foot or so underground to deter rabbits from digging beneath it.

2. IF DEER THINK YOU ARE THEIR PERSONAL SALAD-BAR PROVIDER: Deer will eat almost anything, and to add insult to injury, they'll look at you with those big eyes and bring up all the Bambi guilt. I have met so many people who have simply thrown in the gardening towel, with the explanation "I have deer in my backyard." I have news for you: You can have a garden even if you have deer. Bite the bullet and install a deer fence (make it 8 feet tall, as deer can jump over anything shorter than that). This does mean having your vegetable garden all in one place rather than scattered around your property, but the effect can be very charming if you add some wood posts and a country-cottage gate. You can then plant flowers that deer (and rabbits) typically don't eat—such as rudbeckia, daisies, and echinacea—around the outside border to soften the appearance of the fence (and attract lots of welcome pollinators).

3. IF CHIPMUNKS, SQUIRRELS, AND BIRDS (AND EVERYTHING ELSE UNDER THE SUN) TREAT YOUR GARDEN AS THEIR ONE-STOP SHOP FOR SNACKS: A simple solution to invasion by a wide range of critters is to cover your crops with a row cover, which is a fabric designed to prevent plants from frost and insects—just be sure to remove it once flowers appear on crops that need pollination—or to cover them with mesh netting. Sunflowers and beans just need to make it to a certain size before the critters will leave them alone. If burrowing animals are your challenge, attaching chicken wire to the underside of raised beds will help keep them out.

WEEDS

While animals can be physically barred from the garden to some extent, weeds will find their way in by means of wind, birds, and other methods out of our control. Weeding the garden is necessary to ensure that crops are able to reach their maximum potential. Many weeds are beneficial for the soil as much as they are a nuisance and an obstacle to growing vegetables, herbs, and flowers. So keep some of the good weeds—like dandelion, yarrow, nettles, plantain, and lamb's quarters—as they are both food and medicine to us, as well as providing certain benefits to the plants around them. As for the others, get them quick while they are small and easy to remove. It is best to hoe and pull weeds early in the day to expose their roots to the sun for long enough to kill the plant by drying out its roots.

Weeds compete with your crops by robbing them of nutrients, water, and sunlight. But weeds are also nature's teacher; they can tell us about the soil, and if there is an excess or deficiency of certain minerals,

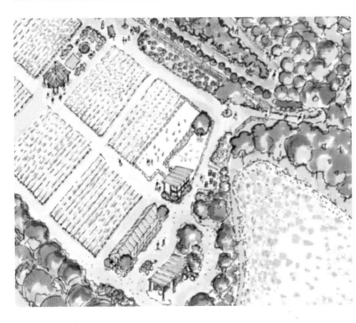

they can help fix it. When you're intentional, you can also turn weeds into fertilizer by incorporating them into your compost, but be sure your compost is getting to temperatures hot enough to kill off any weed seeds or rhizomes that make their way into the pile.

Here are a few tips for combating weeds:

1. THE BEST DEFENSE IS A STRONG OFFENSE. Your first line of attack is to create a healthy, productive garden where weeds don't have room to grow. Plant seeds closer together than seed packages suggest, especially if you plan to harvest often. Use cover crops like crimson clover, even while you are growing other plants, to fill in the spaces and boost the nutrients in your soil. And, finally, mulch, mulch, mulch. In addition to warding off weeds, mulching keeps your soil moist and your plants protected from changes in temperature. Use crushed leaves or wheat straw so that weeds (literally) can't see the light of day. And never underestimate the value of putting newspapers and cardboard in your paths, under wheat straw or wood chips, to keep them weed-free as well.

2. A STITCH IN TIME SAVES NINE. Or shall we say a stirrup hoe to the soil line saves your garden from becoming a weedy mess? (That doesn't rhyme, but you get the point.) Make it a habit to remove weeds often, while they are small, and definitely long before they have a chance to go to seed. You can hand-pull them, hoe them, or even torch them with a crème brûlée torch—just be careful not to hit plants you don't want to lose.

3. A SPOONFUL OF VINEGAR MAKES THE WEEDS GO AWAY. You'll need a solution that's 20 percent vinegar, which can be found at organic gardening supply stores; the vinegar sold at supermarkets typically carries a 5-percent concentration, which is not strong enough to work as a natural herbicide. You'll most likely have to apply this solution a few times, but it's a great way to get rid of weeds that are in cracks and crevices or in the paths and beds where you want to plant. After using the solution, wait at least 24 hours before planting. Some people use boiling water instead of vinegar to do the job. That requires more time to take effect, but it does work, and it's a good choice if you have a conveniently located kitchen garden and want to recycle your hot cooking water.

What about the weeds in your life? Well, the same principles apply. Aim to live a healthy and productive life so that the joys you harvest are bountiful and leave no room for negative energy. Resolve the little problems before they become big ones. And throw hot water (figuratively speaking) on damaging influences before they spread.

GARDEN-VARIETY PROBLEMS?
PUT YOUR BUSINESS SAVVY TO WORK

Garden challenges are great opportunities to boost your business skills. They force you to observe, gather data, utilize your resources, experiment, expand your patience and creativity, and persevere. What's more, garden work lets you learn to live as part of a complex ecosystem, where distinct partnerships can be mutually beneficial. Let's address some common garden problems, and apply some business analogies to solving them:

WRONG PLANT, WRONG PLACE: Doesn't this sound like a human resources problem? Move that completely fish-out-of-water artist from the accounting department to the advertising group, and watch her blossom. Plants are no different. They all have favorite environments for flourishing. If you have a plant that's not happy or productive, research its soil and sun needs and see if it may simply need an internal transfer to another "department" in your garden.

GETTING EATEN ALIVE BY THE COMPETITION: Are all your sales and marketing folks nodding in recognition? If you have a weak product, then your competitors are going to swoop in on you and knock you out. This sounds like what happens with a weak plant. Before you know it, bugs are attacking it, it can't fight off diseases, and it withers on the vine. Your best defense: a strong offense. Build healthy soil that feeds plants the nutrients they need to grow strong, encourage your plants regularly with compost and other soil-boosters, pick off predators and dead leaves frequently, and choose companion plants that naturally deter pests and enhance growing conditions; this will help keep your plants strong enough to ward off predators and diseases.

BIT OFF MORE THAN YOU CAN CHEW? You know that project the boss wants done by the end of the month that has turned into a living, breathing beast that eats up every last minute of your time? Well, if you overextend yourself, your garden can become a project management nightmare as well.

Don't feel bad—sometimes you have to exceed the limit to find out where the limit is. But once you know, scale back, do a small bit well rather than a whole lot poorly, and ask for or hire help when you need it. Just like at the office, use your resources. Shift your perspective to see garden problems as business challenges. You'll use your business smarts to realize solutions for these "garden-variety" problems, and learn new skills in your garden that you can take with you to work. Who knows, you may even nab yourself a promotion from all the growing— literally and figuratively—you're doing in the garden.

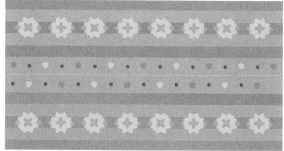

PUT MORE FUN IN YOUR GARDEN

Gardens are inherently living, breathing entities filled with vibrant activity. Why, then, do some of them feel like please-don't-touch museums? If your school or community garden is starting to feel this way, there's a chance you may have beaten the fun out of your garden (or never built it in in the first place). If you haven't heard laughter in a while, or you realize no one is ever out there enjoying the garden, then you may be a quart low on fun. Here are some ways to put some smiles in the aisles:

1. LOOSEN UP. If you've gotten a bit hung up on perfection (perfect beds, perfect rows, perfect image), you may not have left room on the edges for creativity, and the kinds of people (of all ages) who like to have a little fun may not feel comfortable. Now, by "loosen up," I don't mean it needs to be a free-for-all, but perhaps encouraging more open-ended, non-master-planned experimentation will get the creative juices—and people— flowing again. Make sure you truly welcome children as well. Wide paths so they can run a bit, butterfly nets, a place to dig, and kid-size tools all send the message that this is a happy, inviting place.

2. ADD SOME ART. Art personalizes gardens in unique and exciting ways, displays cultural diversity, and provokes conversations that build community. The off-season, when you're not as busy hoeing and growing, is a great time to shift your focus to creating focal points such as:

- Sculptures and statues made by local craftspeople and artisans
- Trellises and ladders, perhaps repurposing urban detritus like old window bars
- "Hardscapes"—walking paths, low stone walls, or mosaic murals using pottery fragments and colored stones
- Living art—select your plantings to create a visually pleasing palette

3. GET CREATIVE WITH YOUR GATHERINGS. If your garden chore assignments are starting to feel ho-hum, try high-energy teams and challenges. Wholeheartedly embrace the people who step forward to lead these kinds of things, and stay upbeat in your communications. Why simply help harvest lettuce when you can have a Halloween "beheading"? Why not bury some hidden treasures for when you ask folks to come help dig holes for fence posts? Why host a basic picnic when you can throw a Fashion Show Fiesta ("Wear your favorite greens!") right there in the garden?

At my garden supply store, we host intimate gatherings in the gardens, often with a bonfire, potluck, live music, and workshops. We bring in local experts who teach about everything from pickling and pruning to composting and canning.

★ TOP 10 ★

EASIEST CROPS TO GROW

*For those with a "brown thumb," or who are just too busy to fuss with finicky crops,
here are a few easy-to-grow favorites:*

1. **RADISHES:** These fast-growing, nutritious little gems are easy to sow directly in the garden. If you want something to build your confidence and break down your belief that you have a brown thumb, grow some radishes. Plant radishes in early spring and late fall, as they like cool weather. My favorite varieties are French breakfast and watermelon radish.

2. **POTATOES:** Have you ever had potatoes sprouting on your kitchen counter? If so, try planting them in the garden, a pot, or even a bucket full of straw. They grow quickly and, unlike seeds, do not need much attention when getting established. Plant in early spring, just after last frost. Try some standard varieties like Kennebec, Red Pontiac, Red Norland, or Yukon Gold to get started. Once you get the hang of it, try growing some of the heirloom varieties like All Blue, Yellow Finn, German Butterball, and Purple Peruvian.

3. **BUSH BEANS:** Growing beans from seed is a breeze. Make a trench 1 or 2 inches deep and sprinkle bean seeds every 1 or 2 inches in the trench. Cover with soil and tamp lightly. If you have a seeder like an Earthway, use the bean plate and set the depth to about 1 inch and go for it. Plant beans after last frost, when the soil temperature reaches 70°F to 80°F. The varieties I recommend for beginners looking for disease resistance and productivity are Provider and Fortex.

4. **GREEN ONIONS:** These are very easy to grow and require minimal effort. Plant green onion sets (bulbs, which will mature faster than seeds) in early spring and fall, as they like cool weather. Simply push the sets in the ground about 2 inches deep, with the pointy sides facing up and nubby sides down. Like potatoes and garlic, green onions do not require much attention until it's time to start harvesting scallions about 30 days later. You can also let them go to bulb, but make sure to pick as soon as you see any of the plants going to flower. You can buy them online or at a local garden center—a collection of red, yellow, and white sets with 100 of each type will typically cost $12.95.

5. **GARLIC:** It doesn't get much easier than garlic when it comes to planting something hardy and productive. Garlic attracts very few pests; in fact, it is a good crop to have in the garden to help deter them. All you need to do is break up a bulb into cloves and plant the cloves about 6 inches apart in rows, with 18 inches between the rows. Push each clove about 2 inches deep into a well-prepared bed, with its nubby side down and pointy side up. Mulch with about 6 inches of shredded leaves, dried grass clippings, and/or straw, and let grow until you come back to harvest 9 months later. You may need to pull a few weeds around the plants in the spring, and it never hurts to add a bit of liquid fertilizer and compost to help get a bigger bulb. A weekly application

in the spring of about 1 tablespoon of liquid seaweed and 1 tablespoon of fish emulsion to 1 gallon of water is ideal. The bulbs are ready to dig up when half to two-thirds of the leaves turn yellowish-brown. Do not water after the beginning of June, to let the bulbs firm up. For beginners I recommend Inchelium Red for a softneck, Music for a porcelain, and Purple Glazer for a hardneck variety.

6. **CHERRY TOMATOES:** Everybody loves to grow tomatoes, but growing big beefsteaks and meaty heirlooms can be harder than it looks. Cherry tomatoes are more forgiving. I recommend starting with transplants, and trellising early to avoid having to get a trellis over or around an already unwieldy plant. Plant soon after the last frost and keep a close eye out for hornworms and blight. Cherry tomatoes do not require much (if any) pruning, making them relatively low-maintenance. My favorite varieties are Black Cherry, Sungold, and Matt's Wild.

7. **ARUGULA:** I can never have too much arugula, but if I do, I like to let it bolt and eat the flowers. It is easy to grow by seeding directly in the garden in spring and fall. Once the plants reach a height of 4 to 6 inches, cut them back about 1 inch above the soil; a second and possibly third cutting will follow. It's best to plant a row or bed of arugula every 2 weeks for a successional harvest. My favorite varieties are Astro and Grazia.

8. **SUNFLOWERS:** Sunflowers are very easy to grow, both for their cut flowers and for their nutritious seeds. They will grow in relatively poor soil but will really thrive in rich, well-drained soil. Plant seeds directly in a well-prepared bed and, so long as you can keep animals from eating their tender shoots, they will grow rapidly and stand strong, with stout stems that can serve as a trellis for cucumbers, pole beans, gourds, and morning glories. The variety I like to grow for edible seeds is Mammoth, which can reach 12 feet in height. For cut flowers, I prefer the pollenless varieties like Sunbright, the ProCut series, and Zebulon for its amazing geometrical pattern.

9. **SUMMER SQUASH AND CUCUMBERS:** Easy to grow from seed, but can be transplanted for a jump-start. Plant after danger of frost has passed and soil temperature is between 70°F and 90°F. Pests such as squash bugs, vine borers, and cucumber beetles can cause problems, but if you are successful, you will be harvesting more than you can eat in no time. Plant radishes around your squash to help deter the vine borers and use Bt regularly (see page 135) to kill them. Summer squash is fine to grow as a bush, whereas cucumbers do best when trellised. The varieties best suited for success are appropriately named: Success PM Straightneck yellow squash, Dario F1 cocozelle squash, Dunja F1 green zucchini, and Marketmore 76 for a slicing cucumber.

10. **MESCLUN MIX:** If you eat lots of salad like I do, then mesclun is great to grow at home. *Mesclun* is a French word for "salad mix"; it is essentially a mix of different lettuces and other greens that are grown close together and cut young for their tender salad leaves. It can be broadcast-sown or direct-seeded in rows placed very close together. I typically broadcast the seed, evenly covering the bed, and very gently rake the seed into the top ¼ inch of the soil. Water gently and often until the crop is ready for harvest, which can be as soon as just 25 days. You can get a few cuttings from each crop, so a successional planting every 2 to 3 weeks is ideal. My favorite is the Organic Mesclun Mix from High Mowing Organic Seeds.

6.

HEALING = COMPASSION

While all produce offers some health benefits, there are many homegrown edibles that have specific healing properties. This chapter will highlight some favorite medicinal herbs and show you how to create a medicine chest from your garden. It will also cover some important lessons on how to care for sick plants, with both preventative and remedial measures. Given my strong commitment to organic practices, I will never suggest using toxic chemicals that could harm your family or the planet.

Gardens and farms can heal in many surprising ways, and it is exciting to see the positive healing effects they are having in places as diverse as corporate headquarters, children's hospitals, prisons, senior centers, and city halls. As you consider the value of your garden, don't overlook the simple and significant healing benefits that it affords you.

Gardens heal in ways that can enable those on the edges of society to feel connected. In this context, healing may mean assuaging feelings of disconnect from society and helping everyone from seniors to incarcerated youth. Gardens can showcase hidden skills and teach new ones, engender a sense of responsibility that gives individuals a reason to keep showing up, and forge healthy relationships and connections across generations in nonthreatening environments.

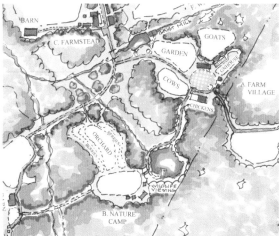

HERBAL MEDITATION

I find growing and working with herbs to be by nature both healing and meditative. The plants themselves exude healing fragrances and energies that put our minds at ease and penetrate to parts of our inner souls. Just working with the herbs, we absorb the powerful healing properties they possess.

Here is a simple meditation for working with herbs. After you have grown, harvested, and dried your herbs, take a large bowl (I prefer wood or metal), a fine-screen colander, and your dried herbs. Find a comfortable place to sit, such as on a back porch overlooking the garden and some bird feeders or in a cozy armchair on a rainy day.

Place the colander over the bowl and begin by rubbing the dried herbs between your hands, letting the rubbed leaves fall into the colander. Gently work the loose herb leaves into the screen by moving your hand back and forth over them, with enough pressure to help the leaves sift through the screen but keep the stems out of the bowl. If you see small stems fall into the bowl, pick them out by hand.

As you rub the herbs between your palms and into the screen, let the healing, fragrant smells ease your mind and relax your body. Remember to breathe, smile, and let go of all your thoughts. Continue rubbing and sifting the herbs into the bowl, and pour the sifted herbs into a clean glass jar for storage. Here are some of my favorite herbs for this meditation:

- SAGE
- MARJORAM
- BASIL
- LAVENDER
- THYME
- ITALIAN OREGANO
- MINT (ESPECIALLY PEPPERMINT)

GARDENS FOR HEALING

There are many healing opportunities that gardens can offer. Getting outdoors and being active helps your body stay flexible and strong. The fresh food from your garden will provide the critical vitamins and minerals you need to thrive. The tea and tinctures you make with herbs can ease your ailments and keep your immune system strong. Here are a few more ways that you can find and share healing in your garden:

1. GARDENS CAN SERVE AS A PLACE FOR REFLECTION FOR THOSE NEEDING COMFORT. People carry their personal worries with them all day long: concerns about an ailing parent or struggling child, financial challenges, or fears about the future. Peaceful moments in gardens, where the sensual joys of plants growing can inspire feelings of hope and serve to counteract those stresses.

2. GARDENS CAN PROVIDE PLANTS THAT HAVE ACTUAL MEDICINAL USES. It's easy to forget that modern medicine originated in plant-based medicinal therapies. The surge of interest in herbal medicine and home remedies is a reminder of the abundant healing benefits of our natural world. Simple herbal possibilities— such as a cup of peppermint tea to energize, some lemon balm to aid digestion, or a sniff of a lavender sprig to encourage sleep—can easily become a welcome new daily habit for those with a small herb garden. More extensive medicinal uses of a wide variety of plants can be learned as well. (Note: Do not consider this medical advice, and tell your doctor if you are using any herbs in addition to prescribed medication.)

3. ORGANICALLY GROWN GARDENS HELP RESTORE THE EARTH. Organic farms and gardens sequester carbon, mitigate water runoff, filter environmental toxins, increase biological diversity, encourage ecosystem development, and enhance the beauty, vitality, and even safety of our communities. Many municipalities are getting into the garden action because of these tangible results, which show that adding a little extra green space enhances the health and welfare of their citizens.

4. ANIMALS ARE HEALERS, TOO. From backyard chickens and rabbits to therapeutic dogs and horses, animals bring a wonderful healing energy to farms and gardens. Cows have a remarkable ability to convert plants they eat into fertilizer that can support their own and our food needs. Cow manure is a biodynamic farmer's black gold and is used in many ways to help restore and replenish the land.

COMMON PESTS AND DISEASES

Of all the things I wish I had studied in college, I think entomology would have been the most useful. Over my many years of farming, I have seen all kinds of bugs, and in most cases I had no idea if they were good, bad, or neutral. I do not like to kill things, but I hate pouring months of hard work, sweat, and tears into a crop and then seeing it destroyed in one night. It is very important to know about pests and diseases, especially those that can wreak havoc on your garden. I recommend having *The Organic Gardener's Handbook of Natural Pest and Disease Control* easily accessible for when something goes wrong in the garden. Here is a brief overview of the most common garden pests and diseases, with some tips on how to deal with them:

APHIDS are pests that attack a wide range of crops, and the best way to know when they are a problem is by using sticky traps. Once you have aphids, act fast by using insecticidal soap (see page 136) or neem (see page 137), and monitor them closely until you have gotten them under control. Remove heavily infested leaves from the garden and burn them or throw them away in the garbage. Introduce predators like ladybugs and green lacewings. It is not recommended to compost pest- or disease-ridden plants unless you are very confident your compost pile is maintaining 135°F temperatures for two weeks or longer.

BACTERIAL SPOT AND BLIGHT are diseases affecting tomatoes, peppers, and many other fruits, flowers, and vegetables. Most bacterial diseases are exacerbated by humid conditions, so it is important to keep good airflow around your plants by pruning and staking, as well as maintaining good irrigation practices to avoid excessive moisture. A biodynamic preventative technique uses a fermented tea of the horsetail plant applied regularly to help increase resistance to air- and soil-borne diseases. There are also some organic controls called biofungicides that use beneficial bacteria, sulfur, copper, and/or neem to fight these harmful diseases.

CABBAGE LOOPERS AND CABBAGE WORMS are the main villains that attack the brassica family; they loop their way into all of our favorite cole crops like broccoli, kale, cabbage, and cauliflower. There are a few ways to deal with these pesky little worms and you will probably need to use all of them. The most effective organic sprays are Bt (see page 135) and Spinosad (see page 137). You can also use floating row cover (see page 136) as a preventative measure, regularly handpick caterpillars, and release trichogramma wasps if you don't have them already present in the garden.

CUCUMBER BEETLES are yellow-bodied flying insects with either black spots or stripes; they feed on the leaves, flowers, and fruit of the cucurbit family, which includes cucumbers, squash, melons, and pumpkins. They also carry and spread the bacterial wilt organism and the cucumber mosaic virus. They look like yellow-green ladybugs but have nothing good to offer your garden, so take action as soon as you spot them. Beetles will overwinter under debris, so keeping the garden free of garden trash is a good practice. As a preventative, securely cover young cucurbit plants with floating row cover until they are big enough that the beetle damage isn't fatal. In a small garden, you can try to handpick the beetles, but your best bet is having a healthy population of beneficial insects such as lacewings, ladybugs, and spined soldier bugs.

CUTWORMS are small caterpillars that can do some serious damage to newly sprouted seedlings and young plants. They overwinter under debris or in the soil and come out in the spring to wreak havoc on crops at night. A fun and effective way to deal with cutworms is to head out into the garden after dark and do some handpicking. Other ways to get rid of them include sprinkling diatomaceous earth (see page 136) around the bases of plants, scattering Bt mixed with cornmeal, and unleashing some trichogramma wasps. Beneficial nematodes are also very effective for attacking and destroying cutworms while they are in the soil.

FLEA BEETLES are one of the most irritating pests in the garden, especially in the spring when you are trying to get everything established. They like cool weather and crusty soil so irrigating, cultivating, and mulching can help. They do the most damage on young plants, so be sure to take measures early to keep seedlings protected. Some techniques include starting from transplants, using floating row cover, planting more seed than is recommended (then thinning once established), and making sure to give seedlings ideal conditions to speed up early growth. Another technique is to plant a sacrificial trap crop such as sweet alyssum or radishes. What I have found works best is a heavy dose of diatomaceous earth applied as a dry powder. Neem and horticultural oil are also somewhat effective liquid controls.

LEAFHOPPERS feed on plants by sucking their juices, and they can transmit viruses in plants such as beans, beets, lettuce, and potatoes. There are over 20,000 species of leafhoppers and they are very common in home gardens. Make sure to keep the garden free of debris and trash to reduce areas for them to overwinter. Use beneficial insects such as ladybugs, lacewings, and pirate bugs to help combat leafhoppers. Floating row covers can be used to help keep leafhoppers off your plants. If they find their way in, thoroughly apply diatomaceous earth and insecticidal soap to reduce the damage.

MEXICAN BEAN BEETLES are a major problem for all types of bean crops. These ladybug-like beetles are yellowish-brown, with sixteen black spots on the backs of their wings. They are most vicious during the summer months, so the earlier you can establish your bean crop, the safer you are. On a small scale, some handpicking can help, and in addition to the bugs themselves, look for their clustered bright-yellow eggs on the undersides of leaves. Similar treatment as for leafhoppers is required to keep them at bay.

POTATO BEETLES are serious pests on potatoes as well as tomatoes, peppers, and eggplants. In the larval stage they are orange with black spots on their sides, and as adults they have a yellowish shell with black stripes and black spots behind their heads. Use beneficial nematodes to attack them while they are still in the soil. A layer of clean straw or hay mulch can also help when applied just before or as soon as plants emerge. In backyard gardens, handpick the beetles and drop them in soapy water. The most effective organic sprays for potato beetles are Spinosad and Bt, which work best if applied when beetles are still young. Beneficial insects are also useful for feeding on the beetles' eggs and larvae.

SLUGS riddle many home gardens. They chew large holes in plants and can destroy seedlings in no time. Be careful with mulch, as slugs like to hide underneath and do their damage at night. Watering in the morning instead of at night will also reduce the population. A good evening exercise is to go into the garden to do some handpicking a few hours after sunset. There are natural ways to trap slugs by using shallow yellow cups of beer, cabbage leaves, or strips of cardboard. There are also a few organic products such as Sluggo, which kills slugs using nontoxic iron phosphate. Diatomaceous earth is another effective natural control for slugs.

SQUASH BUGS can wreak havoc on a crop of summer or winter squash; their damage can be seen in a wilting vine, which often results in the death of the whole plant. It is best to find the bugs' eggs in the garden; they are shiny, slightly oval, and copper-colored and are laid in small masses, usually on the undersides of leaves. Start looking for these as the weather warms and check weekly, making sure to crush any eggs you find.

You will usually find the grayish-brown adult bugs mating on your squash plants—this could mean no butternut squash pies for you, so I recommend compassionately destroying the little lovers. They seek shelter at the bases of plants, so be careful with mulch, as it can help provide a hiding place. I recommend sprinkling diatomaceous earth and sowing radishes and nasturtiums around the bases of the plants to help combat the squash bug.

TOMATO HORNWORMS are highly camouflaged green worms with a black or red horn projecting from the rear; they can do serious damage to tomatoes, peppers, eggplants, and potatoes. It can be easier to see their black droppings (known as frass) than to see the worms themselves. Once you are able to identify them, handpicking is the most effective way to deal with hornworms in a home garden. Spraying foliage with water can help expose the hornworms, as they will likely move around when sprayed. Beneficial insects and applying Bt and/or Spinosad are also effective ways to deal with tomato hornworms.

WIREWORMS: These tough little worms are yellow to brownish-red and spend their time attacking seeds, roots, bulbs, and tubers. They feed entirely underground and so are hard to find until you notice your plants starting to wilt and die. Once again, beneficial nematodes are a great way to attack and destroy wireworms in the soil. Cultivation also helps expose them to the weather and natural enemies. A good way to trap them in a home garden is by cutting a potato in half, running a wire or stick through the middle, and burying it about an inch deep with the wire sticking up out of the soil so you can easily pull it back up. Check the traps every few days and discard the wireworms they will have attracted.

COMMON PEST AND DISEASE CONTROLS

The real key to preventative pest control rests in having a healthy soil full of beneficial nematodes, fungi, and bacteria and a hearty population of beneficial insects living happily and breeding in the garden. In addition, good crop rotation practices, companion planting, selecting resistant varieties, and timing your crops will all help tremendously.

BACILLUS THURINGIENSIS (Bt) is a naturally occurring bacteria common in many soils around the world and has been adapted into an effective organic pest control. Essentially it works by reacting with the cells in the guts of certain insects, leading to paralysis of the digestive system and causing the insects to die of starvation within a few days. It is primarily used for caterpillars such as cabbage loopers, but strains have been developed that are also effective at killing Colorado potato beetles and the larvae of mosquitoes, black flies, and fungus gnats.

BENEFICIAL ORGANISMS are a wonderful natural way to use certain insects for controlling pests and disease. While they are harmful to unwanted pests, they do not harm people, plants, or pets. Some of the most popular beneficial insects are ladybugs, praying mantis, green lacewings, and trichogramma wasps. Others include orchard mason bees, parasitic nematodes, ground beetles, spiders, hover flies, and predatory mites. To get the best results from these organic farming allies, it is recommended that you apply them when pest counts are low to medium. Introducing beneficial insects is not a quick fix, so if you have a pest infestation in your garden, I recommend applying an organic pesticide and introducing the beneficial insects a few days later to take care of the survivors.

Beneficial nematodes are also very effective in controlling harmful insects by attacking them in the soil where they spend much of their time hiding and feasting on the roots of your plants. While good compost provides your soil with many beneficial microorganisms, it is a good idea to introduce those that specifically attack the pests that often wreak havoc in our gardens. There are many great resources for beneficial nematodes and insects (see page 218). These insects can also be attracted to your garden naturally and can become part of a long-term preventative pest solution.

BICARBONATE is pure baking soda and has been shown to be an effective control for powdery mildew. Simply add 1 tablespoon potassium bicarbonate or baking soda and 1 tablespoon of horticultural oil (or 2½ tablespoons vegetable oil) to 1 gallon water. Shake well and add ½ teaspoon biodegradable dish soap (I prefer Dr. Bronner's). Spray the mixture during cool weather on all parts of the plant and surrounding soil every 7 to 10 days until the disease is no longer visible.

COPPER is a natural control for fungal and bacterial diseases, especially on tomatoes. Check product labels to make sure you follow the instructions properly, as this is very potent and application rates are commonly as small as 2 pounds per acre.

DIATOMACEOUS EARTH (DE) is a naturally occurring siliceous rock that consists of fossilized diatoms, a type of hard-shelled algae. It works by drying out the exoskeletons of certain insects, causing them to dehydrate and die. Food-grade DE is also added to cow, goat, and chicken feed as a natural de-wormer. The primary insects that DE is most effective at controlling are slugs, ants, cockroaches, earwigs, flea beetles, and squash bugs.

FLOATING ROW COVER is a white, lightweight fabric made from spun-bonded polyester or polypropylene. While I am not a big fan of using poly products in the garden (such as plastic mulch), floating row cover has some very attractive benefits for the home gardener and small farmer alike. It can function as an effective shield over your plants to protect them from insects as well as birds and animals who want to eat your crop before you do. Do your best to avoid trapping pests inside by covering immediately after you put plants or seeds in the ground and checking for pests regularly. Some other good applications of floating row cover include frost protection and helping young seedlings get established in spring and fall. It is very affordable at about 3 cents per square foot and can last for multiple seasons if you take good care of it.

GARLIC is one of the easiest, most valuable crops you can grow in your home garden; it doesn't take up much space, and it has multiple uses as food, medicine, and pest control. For a water-based garlic spray, use a blender to mix 6 cloves garlic, 1 tablespoon cayenne pepper, 1 tablespoon biodegradable dish soap, and 1 quart water. Pour the mixture into a jar and let steep for 24 hours. Strain and pour the liquid into a spray bottle. This spray will keep for up to 2 weeks in the refrigerator.

INSECTICIDAL SOAP is a potassium- and fatty acid–based soap that disrupts the cell structure of certain insects when applied directly onto the pest. This popular organic insecticide works best on soft-bodied insects such as aphids, thrips, spider mites, mealy bugs, and whiteflies. In combination with horticultural oil, insecticidal soap is also effective against powdery mildew. It is best to do a test spray on a crop, as insecticidal soaps have been known to burn the leaves of certain plants and should not be applied too frequently or during the hottest time of the day.

KAOLIN CLAY is a naturally occurring clay that, when mixed with water and applied to plants, creates a physical barrier that protects vulnerable plant tissue from insects. It is most commonly used in orchards to deter apple maggots, pear psylla, and white apple leafhoppers. It is also relatively effective at combating plum curculio and many types of moths. It should be used in moderation, as research has shown that heavy use can have a negative impact on beneficial insects and cause an increase in certain pests due to the loss of their natural predators. Kaolin clay is also effective at eliminating certain vegetable pests such as onion thrips and pepper weevils.

NEEM is a vegetable oil derived from the neem tree, a drought-resistant tropical tree native to India. Neem oil is a powerful organic insecticide that repels a wide variety of pests, including aphids, cabbageworms, whiteflies, fungus gnats, and leaf miners. Neem is also effective at controlling some common plant diseases such as powdery mildew, black spot, rust, and anthracnose.

OILS DERIVED FROM MINERALS, FISH, AND PLANTS can be used as organic pest and disease controls. These oils can curb a wide range of soft-bodied insects, such as aphids, psyllids, and whiteflies.

PHEROMONES are natural scents created by a female insect in an attempt to attract a male. These species-specific scents are replicated in a laboratory and are used to lure male insects to a sticky glue trap. One or two of these traps in a home orchard will capture the male insects, thereby reducing the rate at which the females can reproduce. These traps are especially effective against certain moths, flies, and beetles.

PYRETHRUM is a plant-based insecticide derived from a chrysanthemum species known as Pyrethrum Daisy—not to be confused with pyrethroids, which are synthetic compounds similar to pyrethrum but not approved for organic production. Avoid pyrethrum products with PBO (piperonyl butoxide), which is a toxic additive that is not allowed under the National Organic Program. Pyrethrum is also toxic to bees, fish, and parasitic wasps, so use sparingly and with extreme caution. It is used to control more than 100 pests and can be used up to day of harvest.

SPINOSAD is made by fermenting a naturally occurring bacterium called *Saccharopoyspora spinosad*. It is a fast-acting organic pesticide that activates insects' nervous systems, resulting in a loss of muscle control and death from exhaustion within just a few days. It has been shown to have relatively low levels of toxicity to mammals, fish, and birds; however, it is toxic to honeybees when wet, so it should not be applied on plants where bees are foraging.

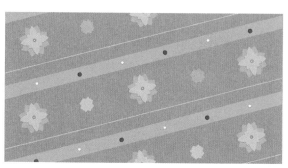

TRAPS are another way to deal with pests. Sticky traps are a great way to see what kind of pests are lurking in your garden so you can address them by whatever natural means possible. Yellow sticky traps will attract a wide range of insects, both good and bad, while blue sticky traps are used for thrips. One of the most effective ways to deal with bugs is by learning their life cycles and focusing on interrupting those cycles with organic sprays or by introducing beneficial predator insects.

WISDOM is your best weapon against pests and disease. The more you know about your enemies, the easier it is to outsmart them. This could be a matter of hitting them hard when they are most vulnerable, catching them before they hatch, or introducing their predators at just the right time.

THE BEST PLANTS FOR ATTRACTING BENEFICIAL INSECTS

Good bugs are your best defense against pests, so make sure to leave room in your garden for some of these amazing pollinator plants. It is a good practice to dedicate at least 5 percent of your garden space to growing flowers that attract beneficial insects. I recommend more like 10 to 15 percent. Some of the following flowers are perennials, which you can get established on the periphery of your garden and in the landscape. Some of the annuals, like sweet alyssum and basil, will bloom for a long time. Others, like buckwheat and coriander, will bloom and die quickly; these are best replanted every 3 or 4 weeks throughout their growing season for maximum benefit.

PERENNIALS

- **ANISE HYSSOP (*Agastache foeniculum*):** This beautiful perennial is easily grown from seed, either direct seeded in the garden or started indoors. It likes full sun and well-drained soil and attracts all kinds of pollinators, including bees, butterflies, and moths.

- **GOLDEN MARGUERITE (*Anthemis tinctoria*):** Also known as oxeye chamomile, this is a clump-forming perennial that thrives in full sun and dry to medium soil. Great for borders in the garden and as a cut flower, this is a powerhouse for attracting a wide variety of beneficial insects.

- **ONION AND GARLIC CHIVES (*Allium schoenoprasum* and *Allium tuberosum*):** These tasty herbs are excellent for attracting bees and butterflies as well as repelling pests like aphids. Both the leaves and the flowers are edible.

- **PARSLEY (*Petroselinum crispum*):** A wonderful edible herb, parsley is also a great plant for attracting beneficial insects such as parasitic wasps and flies. I like to grow both Forest Green Curly and Giant of Italy. Let them go to flower to provide the most benefit in attracting good bugs.

- **WHITE CLOVER (*Trifolium repens*):** This is my favorite perennial pathway crop, especially mixed with a perennial rye or fescue. Clover has so many benefits for both the soil and the atmosphere. As a legume, it fixes nitrogen and enhances the biology in the soil. It is a great food source for bees and also attracts parasitic wasps. It is best planted in early spring, and once established, it provides an excellent pathway between garden beds (as an alternative to grass). It is also a great forage crop for animals like chickens and rabbits and can be established and improved through seasonal overseeding. The clover variety I have had the most success with is Regal Ladino.

- **YARROW (*Achillea millefolium*):** Yarrow attracts beneficial insects during the summer months including minute pirate bugs, parasitic wasps, and beneficial flies.

ANNUALS

- **BASIL (*Ocimum basilicum*):** Basil is a plant of many virtues. Of course we know it best for its culinary value, but it is also a fabulous plant for attracting bees and other beneficial insects. The varieties best suited for this purpose include African, Thai, and holy basil.

- **BORAGE (*Borago officinalis*):** This easy-to-grow annual has edible flowers and leaves and is excellent for attracting bees. It is also a great companion plant for most plants in the garden, especially tomatoes, squash, brassicas, and strawberries.

- **BUCKWHEAT (*Fagopyrum esculentum*):** This is a low-maintenance cover crop that can go from seed to flower in just 3 weeks. It will grow in cool spring weather but is also tolerant of hot, dry weather, making it a great summer cover. In addition to improving soil structure, buckwheat attracts bees, ladybugs, parasitic wasps, minute pirate bugs, and beneficial flies.

- **COSMOS (*Cosmos sulphureus*):** Cosmos is an annual flower that attracts lacewings, hoverflies, and parasitic wasps. Direct seed in the garden after last frost and enjoy blooms all summer long. There are lots of great varieties, my favorites being the Bright Lights blend and White Sensation.

- **DILL (*Anethum graveolens*):** Dill is an amazing plant with many uses; it is a great fresh herb, its seeds are commonly used for pickling, and it is also a remedy for flatulence. In the garden, dill is a favorite food for the Eastern black swallowtail butterfly and attracts aphid midges, hoverflies, lacewings, ladybugs, and parasitic wasps. The Bouquet variety is my favorite; it will self-seed easily so it is best to plant it in an area where it can roam.

- **FENNEL (*Foeniculum vulgare*):** Fennel is a delicious and medicinal herb that attracts butterflies, lacewings, minute pirate bugs, parasitic wasps, and ladybugs. I like Bronze Fennel as a decorative and edible plant that also attracts many of the most beneficial insects. The seeds are also excellent for aiding digestion.

- **SWEET ALYSSUM (*Lobularia maritima*):** This hardy annual is great for borders in the garden and provides a habitat for beneficial predators such as trichogramma wasps and lacewings. These predators can help fight off harmful bugs such as aphids, stinkbugs, and leafworm caterpillars. Sweet alyssum is a great spring and fall flower to support beneficial insects and can be grown in the garden or in containers. Tiny Tim is also a nice variety.

- **ZINNIAS (*Zinnia elegans*):** There are lots of beautiful zinnia varieties to have fun with in the garden. This easy-to-grow annual makes an excellent cut flower and attracts lots of beneficial insects including bees, butterflies, hummingbirds, parasitic wasps, ladybugs, and hover flies. I love the Benary's Giant Series and Oklahoma Formula Mix.

KEEP YOURSELF HEALTHY YEAR-ROUND

I consider any day short of a deep freeze a perfect day for gardening. I love how, after the leaves fall, you can really see the "bones" of a garden—how the light moves across it, and where it might make sense to expand. Typical cold-weather gardening chores including cleaning out garden debris; building and repairing bed frames, trellises, and fences; lubricating and storing tools; planning your spring garden; and just enjoying the simple, stark beauty of a quiet time outdoors. Being out in the garden year-round can also provide you with a full range of health benefits, including:

1. INCREASED HAPPINESS: Both the sun and the soil have been shown to deliver mood-boosting benefits. This can be a welcome effect for everyone, but for those who suffer from seasonal afffective disorder during the winter, it can be life changing.

2. SUSTAINED COMMUNITY CONNECTIONS: You are more apt to see your neighbors during colder months if you are outside working in and around your garden, and these connections can enhance feelings of belonging while reinforcing neighborhood safety.

3. INCREASED AVAILABILITY OF HEALTHY FOOD: Many dark, leafy greens and root vegetables grow during the winter (plus, overwintering crops give you a head start on your spring garden). These seasonal foods are high in immune-boosting properties needed to ward off winter ailments. It's nature's pharmacy, exactly when you need it.

4. INCREASED EXERCISE: It's nice to settle in with a stack of books or a bunch of movies, but a whole winter of that will most likely take a toll on your physique. Weight gain increases risk for heart disease and diabetes, and the weight is also harder to lose the older you get (if you haven't discovered this fun fact yet).

5. REDUCED STRESS: A little time in the garden each day gives you a chance to align your breath with nature's relaxed pace, enjoy simple beauty, and find balance again.

GROW YOUR OWN JUICE

If your day keeps you on the run, you may have discovered that liquid lunches are the way to go—and I'm not talking martinis here. I'm talking kale, carrots, apples, and a whole slew of other healthy goodies pressed through a juicer or perhaps just blended, poured into a travel cup, and taken here, there, and everywhere.

However, it won't take long for you to realize that daily juicing requires a whole lot of inputs. Bags of apples. Piles of carrots. Armfuls of greens. This requires not only frequent trips to the market, but also a whole lot of money. There has to be a better way, doesn't there?

Well, yes, there is. You can grow your own "juice garden" and harvest peak-of-ripeness ingredients that change with the seasons. You'll get a variety of flavors from the convenience of your home for minimal cost. Here's how:

1. PLANT FRUIT TREES AND BUSHES. This is an easy-to-overlook aspect of your garden, as it takes a few years before you get any fruit, but trust me on this. Those few years pass quickly, and the yield you will get from just one fruit tree will keep you up to your eyeballs in juice. Find out what grows best in your climate, and choose varieties that will ripen each month so that you have a continuous supply of fruit.

2. PLANT DARK, LEAFY GREENS. Collards, kale, and Swiss chard all grow large, nutrient-rich leaves that will give your juices a definite power-punch. Adding these leaves to juices that include sweet ingredients like carrots and apples is a clever way of boosting the amount of greens in the diets of anyone (such as kids) who may be a finicky eater.

3. PLANT ROOT CROPS. Did you know that beets have the highest sugar content of any vegetable? Add them to juices, and don't forget to use their greens as well. Other root crops to throw in your juicer include parsnips, rutabagas, kohlrabi, carrots, and sweet potatoes.

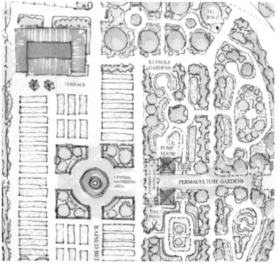

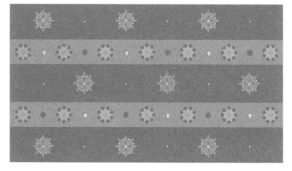

4. PLANT MICROGREENS, SPROUTS, AND WHEATGRASS. No energy or time to go outside and harvest? No space for a garden? No problem. You can grow a garden of greens right on your kitchen counter. Look into starting a tray of microgreens, which are grown from many of the seeds you would plant outside (such as broccoli, cabbage, kale, arugula, and tatsoi), but are snipped and used when they are just a few inches. Sprouts are pretty much the same thing, but consumed at an even earlier stage.

Wheatgrass requires its own special juicer, and you really need only a few ounces of it to get remarkable nutritional benefits; many people either add that small amount to their larger glass of juice or just have a shot glass full of wheatgrass each day. Another benefit of wheatgrass? It is really beautiful when it grows, and it makes a very pretty decoration or centerpiece (as long as it gets enough sun).

5. PLANT PERENNIALS AND HIGH-WATER-CONTENT CROPS. Herbs such as sorrel, mint, and lemon balm provide seemingly endless growth and add a nice variety of flavor along with additional nutrients to your daily juices. "Watery" crops such as celery and tomatoes add abundant liquid to your juice in addition to their nutrients and flavor.

6. USE YOUR BOUNTY. If you already have a robust garden, you know there are times when you are knee-deep in some crop that you can't eat fast enough. Adding juicing to your repertoire (which may also include freezing, canning, dehydrating, and donating) will use up—and preserve—your bounty during times of plenty. Have fun experimenting and coming up with your own beloved flavor and nutrient combinations.

THE GARDENER'S MEDICINE CHEST

The following list highlights some medicinal plants that are must-haves in the home garden. You may never get sick again with all the nutritious vegetables you will be growing and eating at home, but if you do find yourself ailing, try creating your own remedies from these powerful healing plants:

ECHINACEA (*Echinacea angustifolia*): This gorgeous perennial is the best all-around immune booster. It can be grown from seed or from transplants and is hardy even in the coldest of climates. It likes full sun and dry feet, so be sure to plant it in well-drained soil or raised beds. It will self-seed if you let it, which I recommend—you can never have too many echinacea plants in the garden and they make great gifts. While all parts of the plant can be used to make tea or tincture, I recommend digging three-year-old roots in the fall or winter for optimum potency.

PEPPERMINT (*Mentha x piperita*): A potent perennial tea herb with medicinal properties for stimulating digestion and calming the nerves. It is easily propagated through cuttings but should be confined to particular areas in the garden, as it can be somewhat invasive. While it is primarily used for tea, peppermint can also be used as an ingredient in herbal products like lotions, soaps, and toothpaste.

CHAMOMILE (*Matricaria recutita*): This delicate flowering herb is easy to grow from seed and is an excellent tea herb for helping with digestion and relaxation. Chamomile can also be used in herbal eye pillows and skin-care products.

VALERIAN (*Valeriana officinalis*): A powerful perennial herb that is a sedative and antispasmodic and is often used to aid sleep. Valerian is also known to lessen anxiety and nervous tension, relieve muscle spasms, and even help people quit smoking.

GARLIC (*Allium sativum*): Considered both a food and a medicinal herb for its many healing properties, garlic can help reduce heart disease, lower LDL cholesterol levels, and slightly lower blood pressure. It is also a natural antibiotic and will kill many strains of viruses, bacteria, fungi, and parasites.

CAYENNE PEPPER (*Capsicum annuum*): While cayenne is a popular spice, it is also a very powerful medicinal plant. Its main medicinal properties are derived from a chemical called capsaicin, which is what gives peppers their heat; the hotter the pepper, the more capsaicin it contains. Cayenne is used to treat everything from sore throats and headaches to poor digestion and heart disease. It is excellent for stimulating blood circulation. It is easy to grow and a little can go a long way, so just a few plants in the garden should suffice.

SAGE (*Salvia officinalis*): This beautiful perennial herb is often used in cooking and can be used as a gargle to relieve a sore throat. I like to use sage in a pot of beans both because it tastes great and because it helps relieve gas by relaxing the gut muscles. Sage can be grown from seed or transplants and prefers full sun and well-drained soil. It is an evergreen perennial shrub, so once it is established you can enjoy it for years to come, making it a great landscape plant and a must in the herb garden. My favorite variety is common sage for culinary and white sage for ceremony.

THYME (*Thymus vulgaris*): While known primarily for adding flavor to all kinds of culinary dishes, thyme has strong antimicrobial properties. I like to use thyme tea and oil for chest coughs and to promote perspiration at the start of a cold. This is another perennial herb that prefers to grow in full sun and dry, well-drained soil. Sow directly in the ground in early spring or transplant from early summer through fall.

CALENDULA (*Calendula officinalis*): A flowering annual that has both edible and medicinal properties. The flower petals are edible and make a nice alternative to saffron; the flowers are used medicinally for their anti-viral and anti-inflammatory properties. Creams, salves, and other skin treatments often contain calendula as a

key ingredient. It is easy to grow from seed by planting directly in the ground in early spring. While calendula will grow in poor soil and partial shade, it prefers full sun, fertile soil, and cool weather. My favorite medicinal variety is Resina, valuable for its high resin content.

COMFREY (*Symphytum officinale*): A robust perennial that has great value in the garden as both a fertilizer and a medicinal plant. Due to its rapid growth, comfrey does best when grown with plenty of compost. It has a unique ability to mine nutrients from deep in the soil and store them in its large, fast-growing leaves. Medicinally it is very potent and should be used sparingly, primarily as a topical treatment for wounds, bruises, and sore muscles.

DANDELION (*Taraxacum officinale*): Dandelion is best known as a diuretic for treating infections and bile and liver problems. It is usually consumed as a tea; both the leaves and flower buds are edible and lose their bitterness when cooked. Dandelion roots can also be roasted and used to make a caffeine-free coffee beverage, and the flowers can be brewed into dandelion wine. I like to add the greens raw to a cabbage salad for a little extra kick. While the wild variety is just fine for me, there are a few varieties that are more productive and slower to bolt, such as Clio, Red Rib, and Catalogna Special.

WHITE YARROW (*Achillea millefolium*): A hardy, drought-tolerant perennial herb with a wide range of healing properties—as a decongestant, as a blood coagulant, and for reducing inflammation. Yarrow is usually used as a tea, and also topically to stop bleeding. Yarrow can be started from seed or divisions and likes well-drained soil in full sun, though it will tolerate some shade and a wide variety of soils.

LEMON BALM (*Melissa officinalis*): A very easy-to-grow perennial and wonderfully calming herb that controls high blood pressure, depression, and migraines. It spreads like wildfire, so you will have plenty to share with friends. In addition to being used as a medicinal tea for calming the nerves, it is a good mosquito repellant. Simply crush the leaves and rub on your skin.

STINGING NETTLE (*Urtica dioica*): Sadly, many people try to get rid of this magical plant because of its sting when you touch it. It is a medicinal superpower, with a long list of healing properties: as a treatment for allergies, bladder infections, hair loss, and celiac disease and as an excellent blood purifier. It has diuretic properties that alkalize the blood and release uric acid from joints, so it is a great herb for people suffering from gout or arthritis. It is very high in iron and helps combat anemia. It is also excellent for women, as it supports the liver and the female hormonal system; it promotes milk production in nursing mothers, relieves menopausal symptoms, and reduces excess menstrual flow. It helps break down kidney stones and reduces an enlarged prostate.

Stinging nettle is also great for skin issues such as eczema and acne; it can help remove warts when used topically. It is often found as an ingredient in shampoos, as it has a stimulating effect on the scalp; it can be used to relieve dandruff and help regenerate hair growth.

ST. JOHN'S WORT (*Hypericum perforatum*): This pretty perennial herb is best known for its ability to treat depression. It also doesn't mind growing in poor soil. It is very easy to grow and can be propagated from either seed or cuttings, preferably in the spring or fall.

GARLIC AS MEDICINE

Here are four great ways to use garlic medicinally:

1. GARLIC TOAST TWO WAYS: *Rubbed*—Cut a clove of garlic in half lengthwise and rub the face of the garlic onto hot toast. A little drizzle of olive oil and sprinkle of salt will make for a quick and easy garlic toast with no stinky blenders to clean up afterward. *Blended*—Pour olive oil in the blender up to the top of the blade; add a few cloves of homegrown garlic and a large pinch of natural sea salt. Crank the blender up to high, then brush the garlic oil on bread or toast.

2. GARLIC, LEMON, AND CAYENNE "COLD KILLER": This is my first line of defense if I feel myself getting sick. It is a powerful immune booster and cleanser that tastes better than it sounds. Add the following organic ingredients to a quart of good-quality water: 1 minced clove of garlic, the juice of 1 lemon, and a sprinkling of cayenne pepper to taste (you want to feel the heat). If you want to add a little more of a boost, add a chunk of minced and juiced ginger. Oh yeah!

3. GARLIC SHOE TRICK: Getting kids to eat garlic is not so easy, but this is a great little trick for keeping your little ones from getting sick, especially with coughs and colds. Stick a peeled clove of garlic in between your child's sock and shoe (or do the same with your own). You will know it's working when you smell garlic on your child's breath (or taste garlic in your mouth). Another way to do this is to coat the soles of your feet with a little olive oil, then rub smashed garlic on them.

4. GARLIC FOOT BATH: To kick a cold or prevent a new one from moving in, fill a large pot with water and bring to a boil. Meanwhile, peel 10 to 15 cloves of garlic and crush them, using the side of a knife or a mortar and pestle. Pour the boiling water into a basin where you will soak your feet and add the garlic to the boiling water. Allow the water to cool just to the point that you can handle putting your feet in it—the hotter the better, but be careful not to burn yourself. Soak your feet until the water becomes cool, then either rewarm and soak some more or dry your feet.

"Wear gratitude like a cloak and it will feed every corner of your life."

Rumi

7.

REAPING = GRATITUDE

While the journey is the destination with gardening, the rewards of the harvest are oh-so-sweet: delicious, nutritious food and healing medicine for family and friends, closing the gap between you and your sustenance, and enjoying the fun and pride of growing from within.

In organic farming and gardening, nothing goes to waste. The best-quality produce is sold, the seconds are often eaten by the farmer or donated, and the rest ends up in the compost pile, returned to the soil to help grow more food. Nature is perfect, and we should remember to be grateful for the abundance it provides. The sun rises and sets, rains fall, seeds grow, and we are nourished and sustained day after day, year after year, and generation after generation. Our home, Planet Earth, has been dramatically affected by the way our lifestyles invade these fragile processes. When we are aware and grateful for the natural gifts that sustain us, it is more likely that we will work to protect and preserve them.

With a garden, there are always going to be some failures. One important lesson I have learned from mentors along my journey is to celebrate your accomplishments whenever possible, as it is easy to get swept up in the next task and forget to be grateful for your successes—big or small.

In this chapter we will reap a little bit of everything, from salad mix to squash blossoms; we will learn when and how to harvest and what to do with the bounty. In addition to harvesting techniques, I will explore ways for you to get maximum culinary enjoyment from your garden with some favorite farm-to-table recipes I have enjoyed with family and friends at my home or theirs, inspired by my own growing experiences.

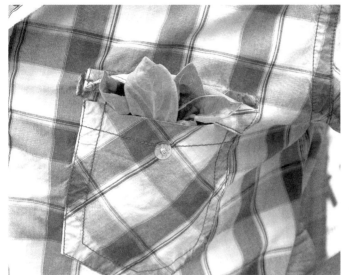

★

HARVEST MEDITATION

It is with overflowing thanks that I approach any harvest. It is not ironic that we often bow down toward the earth when picking produce. I see this as a gesture of humble gratitude to the forces of nature for conspiring perfectly to grow these seeds to fruition.

The harvest meditation involves a very simple, Zen approach. Go out into your bean or pea patch with a harvest bucket or basket in hand. Give thanks for the bounty and begin picking the ripe, plump pods ready for harvest. Be gentle but firm with the plants as you reap, intuitively sensing which ones to harvest and which to leave for a later date. This process relaxes the mind; in fact, I find that the less I think, the more quickly and enjoyably it goes. Be present, be gentle, and give thanks as you pluck away at these living plants. Let yourself feel the breeze, hear the insects and birds, absorb the sunshine, and appreciate the elements that have helped you reach this point.

Some of my favorite harvest meditation crops are:

- BEANS (ALL TYPES—BUSH, POLE, SOY, DRY)
- PEAS (ALL TYPES—SHELLING, SNAP, SOUTHERN)
- CHERRY TOMATOES
- KALE, COLLARDS, CHARD, MUSTARD
- CUCUMBERS
- SUMMER SQUASH
- ROOT VEGGIES (RADISH, TURNIP, CARROT, GARLIC)

★

TOOLS FOR REAPING

HARVEST KNIVES: A good knife is a gardener's best friend, whether it's for cutting twine, opening a bag of feed, or harvesting a ripe veggie. I like to have a good pocketknife on me at all times; I prefer one that is not serrated and has a colorful handle so I don't lose it in the garden. A good pocketknife will cost between $10 and $30. I also like to keep a field knife and a machete in the tool shed for harvesting thick-stemmed crops like cabbage, broccoli, and cauliflower. A 6- to 8-inch field knife and a 14-inch machete with high-carbon steel blades will cost from $15 to $25 each.

HARVEST BASKETS: When it's time to go out and harvest the evening's dinner, what will you use? Rather than grabbing a shopping bag or kitchen bowl, celebrate the harvest with a handmade harvest basket. These can usually be found at a crafts market and often present an opportunity to support local artisans. Another popular option is the wood-and-wire Garden Hod harvest basket from Maine Garden Products, which costs about $40.

SALAD SPINNER: A basic home salad spinner will do for a few rows of salad mix, but if you get a little more productive and need something bigger, check out Dynamic salad spinners (costing $90 and up).

PISTOL NOZZLE: When you're cleaning root veggies, it helps to have a high-pressure nozzle for knocking the dirt off your carrots, beets, turnips, and rutabagas. There are lots of options ranging in price from $10 to $20.

GARDEN CART: While wheelbarrows work pretty well for moving heavy stuff around the garden, nothing replaces a good old-fashioned garden cart. There are two main options here. One is a folding aluminum garden cart for about $250, which is great for people with limited space. The other is a classic wooden garden cart such as those made by Carts Vermont; a medium-size cart will cost from $250 to $300 and a large one from $340 to $400. These may be overkill for a backyard garden, but boy, they sure are nice!

POINTED SHOVEL: A sharp, pointed shovel is useful for harvesting root crops such as carrots, garlic, potatoes, and sweet potatoes. Harvesting a long row can put some wear and tear on both you and the shovel so I recommend purchasing a heavy-duty, well-made model.

TIPS FOR KNOWING WHEN AND WHAT TO HARVEST

One of my great joys is visiting school or community gardens that I had a hand in helping start and seeing how things are growing. What I notice, however, is that many crops become overmature before being harvested. By waiting too long, you may not be enjoying the fruits of your labor at peak ripeness, you may not be maximizing your plants' yields, and you may actually be inviting unwanted pests into the garden by letting food spoil.

Hang around a community garden or volunteer on an organic CSA farm and ask questions. You will learn all the ins and outs of each crop in no time at all, plus how to use more parts of the plant than you may have thought possible. Track the time between planting and harvest, as this varies by crop and variety. Here are some basic guidelines for timely harvesting:

BEANS, PEAS, AND CUCUMBERS: The more you pick, the more you get. Pick often and the plants will "get the message" that they should keep producing.

BROCCOLI: It breaks my heart when I see people have waited too long to pick broccoli; the big heads, when left too long, will open in a spray of yellow flowers. You want to cut off the heads before this happens. You won't get another head on the same plant, but you will send the message to the plant to start producing broccoli side shoots, or florets.

HERBS: Snip off flowering tips on herbs to encourage the plants to become more bushy and keep producing.

LETTUCES: You have three choices here. If your lettuce is a variety that forms heads (such as buttercrunch), you can leave it alone until it forms a nice, tight head and then cut it off at the soil line, but then it's gone. For all types of lettuce (including head lettuce), you could "cut and come again," which means you snip off the outside leaves and leave the center to keep growing, which works particularly well in a home garden. You could also "broadcast" lettuce seeds by tossing them very closely together when you plant, and then "haircut" the baby lettuce greens as they grow by grabbing a bunch in your hands and just snipping straight across with sharp scissors 1–2 inches above the ground.

MELONS: Muskmelons and cantaloupes will slip off the vine with just a slight tug when they are ready. Watermelons, on the other hand, will not. They will have a yellow belly, a brown "pig's tail," and a hollow thumping sound when they are ready. (Hint: The "pig's tail" is the curly tendril on the vine closest to the watermelon. You'll see what I mean when you grow one!)

OKRA: Cut pods with scissors or kinfe every few days, before they are longer than 2–4 inches.

SUMMER SQUASHES: Pick small and often. Although it is tempting to grow the biggest zucchini possible, you really want to harvest summer squashes like zucchini, pattypan, and yellow squash when they are small to medium, for the best flavor and texture. Pick male flowers (on the ends of the stems, not fruits) for squash blossoms.

See Top Ten lists in each chapter for more harvest tips.

GROWING AND USING CULINARY HERBS

You know the feeling. You have four burners going and the oven on as you're preparing dinner for eight, when you stop short at the sight of the next essential ingredient in your recipe: the distinct flavor of chopped fresh tarragon. Not only do you not have it, but you're not in a position to run out to the store and get it (if the store would even have it). And considering the high price of herbs, you start to think there must be a better way.

Luckily, there is. Herbs are easy to grow, many of them are perennial, and they can be used in a variety of ways. If you love creating interesting dishes in the kitchen, you'll enjoy growing herb varieties that you can't find in a store, such as cinnamon basil and chocolate mint. What's more, many herbs can grow in small spaces, such as in a raised garden bed or pot right outside your kitchen door, so all you need to do is step out and snip. Here are some tips for making the most of your herbs:

1. PLANT YOUR HERBS STRATEGICALLY. Some herbs, like rosemary, lavender, and sage, really like to stretch their arms, so these would be best planted in spots where they will have room to grow. Others, like thyme, like to creep and crawl and look especially nice hanging over the edges of rocks or a raised garden bed. Mint and lemon balm spread like wildfire, so they work best in a pot or raised garden bed. Keep herbs like dill, oregano, basil, and cilantro trimmed regularly, and they can live happily together in a relatively small space.

2. FEED YOUR HERBS OCCASIONALLY. Just as you enjoy good food, so do plants. Their delectables, however, are compost and organic fertilizers. Herbs are notoriously low-maintenance, however, so be careful not to overdo it. Annuals like basil seem to benefit from more frequent care. Rosemary, lavender, and sage are fairly independent and aside from pruning them for harvest, they can pretty much be left alone.

3. SHOWCASE YOUR HERBS FREQUENTLY. Offer your guests the opportunity to select their own herbs as toppings for their salads. Let overnight visitors choose their morning tea leaves, and brew a fresh cup with herbs like chamomile and lemon balm. Wrap rosemary sprigs with twine onto gifts, and lay lavender on a pillow to encourage a good night's sleep. Clip four-leaf clusters of basil to punctuate a pasta dish, or tuck some refreshing peppermint into school lunch boxes.

4. DRY YOUR HERBS. Place herbs in a dehydrator overnight (or simply hang them upside down in a closet for two weeks) to make tea, culinary spices, bathing salts, and more. These herbal creations make great gifts and are nice just to have around to add a little extra spice to your life.

THREE WAYS TO GET FAST FOOD FROM YOUR GARDEN

My entire business revolves around showing individuals, communities, and companies how to grow their own food. But like many of you, my days require long hours, lots of travel, and many meals in restaurants or conference rooms. Often, I arrive home after dark, and those precious moments I like to spend in my garden or preparing fresh food don't always happen. Especially with an energetic toddler in tow, both my wife and I are pretty much stretched to the limit.

Is there room in this lifestyle for cooking from scratch with ingredients we grow ourselves? Not only is there room, there is necessity. If there's one thing I've learned through the years, it's that getting proper nutrition on a daily basis is essential for maintaining the pace required for my life and business. I've also learned that honoring my food is important to me, and I've developed some tricks for making it easy to do so. Want fast food that's also fresh and healthy? Here are some ideas that might fit your busy lifestyle:

1. TAKE TWO HOURS EACH WEEK TO PROCESS. Let your garden know you mean business. Slice, dice, sauté, puree, bake, and roast your weekly bounty. You won't believe how much you can get done in this time frame. This small window dedicated to processing your homegrown food could become your favorite hours of the week. It's a time to be quiet, let light stream in the window, and honor the fruits (and veggies) of your labor.

2. FILL THE FRIDGE. Make your newly processed food easy to "grab and go" by packaging it for the realities of your life. Put sliced raw fruits and veggies in small containers to have in the car for your commute; you'll find yourself snacking on red pepper strips and strawberries rather than that chocolate bar. Make a sauce with roasted tomatoes, eggplant, garlic, zucchini, and even butternut squash, and precook some pasta, so all you have to do is toss and heat when you come home. Pick, wash, and chop enough fresh greens for a few days, and prepare an herb dressing so your salad fixings are right at your fingertips.

3. FREEZE FOR LATER. Make pesto from all that wonderful fresh basil and freeze it in ice cube trays so you have it to add to soups, sauces, and pasta dishes, or simply to spread on warm bread on that first chilly day of fall. Freeze sauces in 1-cup batches so you can defrost just what you need. Shred that bumper crop of squash in seconds in your food processor and freeze it in batches that fit your favorite zucchini bread recipe, so that making healthy muffins for a quick breakfast or a midday snack is a snap all winter long. Make at least one dish a week in which you can throw as much of your garden harvest as possible—think rice dishes, soups, quiches, and vegetable pancakes. You can vary these by changing the flavorings and the grains (such as farro, quinoa, or lentils) and, of course, the herbs and veggies will vary naturally with the seasons.

★ TOP 10 ★

EDIBLE FRUIT AND NUT TREES

Want to plant some things that will feed your grandchildren? Plant fruit and nut trees throughout your yard to provide food, shade, and wildlife habitat. All trees are best planted in winter, when they are dormant, or very early spring before the heat of summer kicks in. There are dozens of varieties to choose from and so I recommend referring to your local nursery or university extension to find out which varieties do best in your climate. Here are some popular backyard fruit and nut trees that can produce for decades:

1. **FIG TREES:** A great option for the home garden, as they are pretty low-maintenance. My favorite varieties are Celeste, LSU Purple, Alma, and Brown Turkey. My wife and I planted a tree as part of our wedding ceremony, and we went with a semi-dwarf fig variety called Blackjack, which is great for urban environments.

2. **AMERICAN PERSIMMON TREES:** A delicious and hardy fruit tree that can thrive in many climates. I recommend going with a nonastringent variety liked Fuyugaki or Giant Fuyu.

3. **PAWPAW TREES:** Cold-hardy trees that produce soft, custardy fruits, which can get as big as 2 pounds each. They require more than one tree for pollination, so be sure to plant a few if you want to get fruit. Plant in winter and harvest in fall.

4. **DWARF APPLE TREES:** Cold hardy and very popular for backyard and front-yard gardens. Apples will grow in a wide variety of climates and there are many varieties to choose from. For the best results and optimum disease resistance, I recommend sourcing varieties that do best where you live.

5. **ASIAN PEAR TREES:** Well adapted to warmer climates. Some popular varieties include Chojuro, Hosui, and Shinko. They do best with another pollinator, so plant in pairs—no pun intended.

6. **NECTARINE TREES:** These small trees are well adapted to urban environments. They are similar to peaches but tend to be a bit easier to grow. Nectarines will grow in Zones 5 to 9, and I recommend checking with local nurseries to see what varieties do best in your area.

7. **CITRUS TREES:** If you live in a citrus-friendly place like California or Florida, you have lots of options to choose from—lemons, limes, grapefruit, oranges, and many more. If you are not in Zone 10, then you can try to plant them in large pots and bring them inside a sunny room for the winter.

8. **FILBERT TREES:** Also known as hazelnut trees, these are great nut trees for the home garden and require very little maintenance. They will grow in Zones 5 to 9 and are self-pollinating, so they can produce even with just one tree. Make sure they have plenty of room to spread out, as they will get 20 to 30 feet tall.

9. **CHESTNUT TREES:** Wonderful trees that produce an abundance of nutritious nuts. The American Chestnut was almost wiped out completely by a chestnut blight in the early 1900s. While you can still plant the hardier American Chestnut, the Chinese Chestnut seems to be more reliable as a food producer for backyard gardens. Chinese Chestnuts will survive in Zones 5 to 10, require two trees for pollination, and prefer sandy soils and full sun. Make sure you have plenty of room, as they will grow up to 60 feet and will shade out to about a 40-foot diameter.

10. **POMEGRANATE TREES:** These are hardy Mediterranean trees that typically prefer a warm, dry climate. There is, however, a Russian Pomegranate that can withstand temperatures into the single digits. They grow 15 to 30 feet tall, and though they are self-pollinating, they do best planted in clusters. Pomegranates have great nutritional value and are loaded with vitamin C.

★ TOP 10 ★

CROPS FOR CUTTING GROCERY BILLS

Who doesn't want their garden to pay for itself and then some? Here are some crops that can save you real money at the grocery store, and maybe even make you a few bucks:

1. **SWEET POTATOES:** These are easy to grow, require very little expense or maintenance, and are possibly the most nutritious vegetables in the garden. In lieu of expensive transplants, sweet potato plants come as "slips," which cost between 7 and 25 cents, depending on how many you buy at a time. They also do not require your most fertile soil and can be grown on fences and down hills where other crops can't. The best part is that they produce about 2½ pounds per plant, and at a retail cost of about $2 per pound, versus the cost of about 50 cents to grow, you are looking at a profit of about $4.50 a plant. That is about a 900 percent return on your investment! Don't forget you can eat the leaves, too.

2. **ZUCCHINI:** This plant is a prolific producer of food, with an average yield of about 5 pounds per plant and the potential for up to 10 pounds! That's a lot of food from one seed. At an average value of about $2.50 per pound retail, that's a gross of around $25 per plant. Your cost is at most $2 for seed, fertilizer, and labor. That is an impressive profit, if you ask me. My favorite varieties are Costata Romanesco and Raven.

3. **CUCUMBERS:** Cukes are similar to zucchinis in that once they are established, they need to be picked every few days to prevent their fruits from getting too big for eating. The big difference is that cucumbers fare much better with a trellis. A typical slicing cucumber can produce about 15 fruits per plant, which will sell for about $2 apiece, which is a $30 yield for a $1 investment. Not bad! A 3-by-8-foot bed can produce up to 70 pounds of cucumbers. At about $3.50 per pound, that is $245 worth of cucumbers in a single 24-square-foot raised bed. My favorite high-yield varieties include Marketmore 76, General Lee, and Suyo Long.

4. **HEIRLOOM TOMATOES:** This is probably the most popular—and expensive—organic crop sold at markets— an organic heirloom tomato can cost as much as $7 per pound. While they are not easy to grow, the rewards are well worth it. One well-tended tomato plant can yield up to 10 pounds, and at even a conservative $5 per pound, that's $50 worth of tomatoes per plant. The cost to grow a tomato plant is closer to about $15 when you factor in buying a transplant, compost, fertilizer, a trellis, and labor as well as pest and disease controls. Some of my favorite high-value heirloom varieties include Cherokee Purple, Black Cherry, Green Zebra, Brandywine, Paul Robeson, and Mortgage Lifter.

5. **BELL PEPPERS:** These are also some of the most expensive veggies to buy at a market, with prices as high as $5 per pound. This is largely due to the fact that they are not easy to grow, but if you have success, you will reap the benefits of some major savings. Red peppers are also one of the top twelve crops to buy organic, as they are heavily sprayed when grown by conventional methods (see Dirty Dozen list on page 164). A homegrown bell pepper plant can yield as much as 8 pounds per plant, which is about $25 gross, based on a conservative price of $3 per pound. The cost to produce bell peppers is roughly $5 for the transplant, fertilizer, trellis, labor, and pest and disease controls. Still a great value! The varieties I recommend are California Wonder, Ace, Carmen, and Ozark Giant.

6. **EGGPLANT:** This is another crop that can produce good value for the home gardener—if the flea beetles can be kept at bay. A well-tended Japanese eggplant crop can yield about 15 fruits per plant, and a more classic

bell type will yield about 5 pounds per plant. Organic eggplant at the grocery store is going to run around $2.50 per pound, so you are looking at a value of about $12.50 per plant. Though not as profitable as some others, the cost to grow eggplant is about $5 per plant, so you are still more than doubling your investment. The varieties I recommend for the best yields are Black Beauty and Orient Express.

7. GARLIC: When grown organically, garlic can sell for as much as $16 per pound. While it takes up space in the garden for 9 months, most of those months are during the winter and you can produce a lot of garlic in a small space. You can expect about ⅛ of a pound per square foot, which means a 5-by-10-foot bed of garlic can yield about 6 pounds of garlic, at $10 per pound, which is a value of $60. Your investment: 1 pound of garlic seed at about $10, plus the time it takes to plant and harvest. If you want a highly productive type that will turn 1 pound of seed into 12 pounds of bulbs, go with an artichoke variety like Inchelium Red, Polish White, or California Early.

8. LETTUCE: Lettuce is a very rewarding crop for a backyard garden, as there are few things that taste better than a fresh head of butter lettuce. A typical $4 packet of lettuce seeds will contain upwards of 500 seeds. If half of them grow, you have 250 heads of lettuce (if you have the room) at $3 a head, which is $750 worth of lettuce from a $4 packet of seeds. You do have some compost, fertilizer, and labor costs to factor in, but you have a serious return on your investment here. Plant a new area every 2 weeks during the lettuce-growing season to ensure you have an ongoing supply. Some of the popular heavy-yielding types are Winter Density Bibb, Ermosa, Buttercrunch, Crispino iceberg, Sylvesta butterhead, Tropicana Leaf Lettuce, New Red Fire, and oh-so-many more you can try.

9. BROCCOLI: This may be my favorite vegetable and we eat a lot of it in my house—more than we can grow in our backyard, unfortunately. It is not easy or quick to grow, and it's pricey, too, but worth the effort. A head of organic broccoli at the grocery store goes for about $3, and if you grow it yourself, a packet of 100 seeds will cost about $3.50. You can reasonably produce 25 heads out of that packet in a 3-by-8-foot raised bed, which would be valued at about $75. It doesn't make as much financial sense if you are buying transplants at $3 per plant, which is a good example of why it's so valuable to get good at starting crops like this from seed. My favorite broccoli varieties are De Cicco, Arcadia, and Marathon. If you are feeling very adventurous, try growing a few Romanesco broccolis, as their cool fractal shape makes for a great conversation piece in the garden and on the table.

10. WINTER SQUASH: There are many varieties, and most of them have amazing storage life and nutritional quality. Let's take butternut, probably the most popular of all the winter squashes, which goes for about $1.50 per pound, with each fruit weighing roughly 3 pounds. That's a cost of about $4.50 per fruit. A packet of about 30 seeds will cost around $3, and each plant should yield about 4 fruits. That means if half of your seeds grow to maturity, then 15 plants will produce about 60 fruits, which is a value of about $270. You may have butternut squash climbing over your whole backyard but you will not go hungry, as 60 butternuts will feed your family and friends all winter long. Most of the winter squash varieties are vining, so they will sprawl out over your garden—with the exception of acorn squash, which stays bushy like a summer squash. Acorns yield 5 to 6 fruits per plant, at about 2 pounds per fruit. My favorite acorn varieties are Tuffy, Table Queen, and Sweet Reba. My other favorite winter squash include Red Kuri, Burgess Buttercup, Waltham Butternut, Delicata, Spaghetti, and Blue Hubbard.

THE DIRTY DOZEN AND THE CLEAN FIFTEEN

The Environmental Working Group tests produce for pesticide content and provides a list of crops to either avoid or buy organic, as well as a list of conventional crops with the lowest pesticide contents (see www.ewg. org). This is especially valuable when you are trying to shop on a budget, so you can know which conventionally grown produce is okay to buy (The Clean Fifteen, below) and which you should avoid (The Dirty Dozen). It's helpful to know when it is worth spending the extra money to protect yourself from harmful organophosphate insecticides.

THE DIRTY DOZEN (PLUS TWO)	THE CLEAN FIFTEEN
1. Apples	1. Onions
2. Celery	2. Sweet corn
3. Sweet bell peppers	3. Pineapples
4. Peaches	4. Avocado
5. Strawberries	5. Cabbage
6. Nectarines (imported)	6. Sweet peas
7. Grapes	7. Asparagus
8. Spinach	8. Mangoes
9. Lettuce	9. Eggplant
10. Cucumbers	10. Kiwi
11. Blueberries (domestic)	11. Cantaloupe (domestic)
12. Potatoes	12. Sweet potatoes
13. Green beans	13. Grapefruit
14. Kale and other greens	14. Watermelon
	15. Mushrooms

FARM NIBBLES

As I graze my way through farms and gardens tasting fresh-picked leaves, fruits, herbs, and flowers, I often make a little wrap with a sorrel, lettuce, or collard leaf and a filling of whatever is looking ripe for the munching. There are no hard-and-fast rules for these concoctions; they are just spontaneous expressions of gratitude to the earth, and I call them Farm Nibbles.

The other time I find myself wrapping, stacking, or stuffing myself a nibble is when I am in the kitchen preparing a meal. These nibbles can get a bit more sophisticated with ingredients like olive oil, balsamic vinegar, salt, cheese, avocado, and other yummy things within arm's reach of the butcher block.

While Farm Nibbles can easily be made by hand in the garden, I suggest carrying a sharp knife with you, giving you some options for slices and dices instead of chunks as you craft your own Farm Nibbles al fresco. Some examples of knife-friendly Farm Nibble crops are cucumbers, radishes, carrots, tomatoes, and squash.

TOP TEN FARM NIBBLE WRAPPERS

1. Lettuce (a buttery leaf is best, but any lettuce leaf will do)
2. Collards (raw or lightly steamed or braised)
3. French sorrel (a wonderful perennial herb with a strong lemony taste)
4. Sweet basil
5. Spinach
6. Arugula
7. Tatsoi (an Asian green with leaves similar to spinach)
8. Mustard greens
9. Chard
10. Kale (lacinato is my favorite variety)

TOP TEN FAVORITE FILLERS

1. Beans
2. Peas and pea tips
3. Tomatoes
4. Cucumbers
5. Radishes, carrots, and turnips
6. Green onions
7. Edible flowers
8. Cheese
9. Avocado
10. Sauerkraut, pickled veggies, and kimchi

FAVORITE FARM NIBBLE CROPS BY SEASON

SPRING	SUMMER	FALL	WINTER
• Lettuces	• Beans	• Okra	• Spinach
• Arugula	• Tomatoes	• Kale	• Carrots
• Green onions	• Peppers	• Collards	• Collards
• Broccoli	• Squash and squash blossoms	• Lettuces	• Kale
• Broccoli rabe	• Cucumbers	• Radishes	• Apples
• Sorrel	• Basil	• Carrots	• Parsnips
• Cilantro	• Dill	• Tomatoes	• Pickled cucumbers
• Radishes	• Fennel and fennel pollen	• Cucumbers	• Pickled okra
• Turnips	• Arugula blossoms	• Apples	• Pickled sweet and hot peppers
• Peas	• Chive blossoms	• Peas	• Dilly beans
• Carrots	• Gem marigolds	• Turnips	• Pickled carrots
• Violas	• Strawberries	• Edible flowers (nasturtiums and violas)	• Jams and jellies
• Nasturtiums	• Blueberries		
• Strawberries	• Figs		

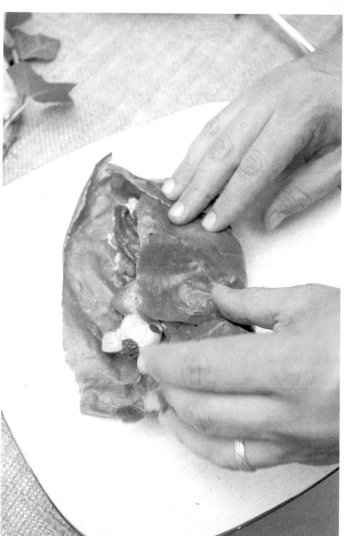

★

A FEW OF MY FAVORITE FARM NIBBLE COMBOS

Unless indicated otherwise, use the leaf to wrap the other ingredients.

BREADLESS BRUSCHETTA

1 large basil leaf

1 tomato slice

1 thin slice green onion

1 small slice of cheese (preferably raw cow, goat, or sheep)

Sprinkle of salt

Drizzle of olive oil and balsamic vinegar

LEMON LEAF WRAP

1 sorrel (lemon) leaf, 4 to 6 inches (10 to 15 cm) long

1 radish, bitten in half or sliced

1 small broccoli floret

2 to 3 edible flower blossoms (borage, violas, nasturtiums, etc.)

SPRING BUTTER-LEAF LETTUCE WRAP

1 large lettuce leaf (butter leaf is my preference)

2 green beans (yellow or purple are fun too—any snap bean really)

1 radish, bitten in half or sliced

2 pea tips and/or snap peas

3 to 4 edible flowers (borage, violas, nasturtiums, etc.)

PEA WRAP

1 large lettuce leaf

Several cherry tomatoes

Several shelled peas

Several edible flowers

Shake of evening primrose seeds (several pods' worth)

LEMON LEAF WRAP WITH PEA TIPS AND VIOLAS

1 sorrel (lemon) leaf, 4 to 6 inches (10 to 15 cm) long

2 pea tips

3 edible violas

"LION AND LAMB" WRAP

(Even if you don't have a garden, you can make a Farm Nibble out of weeds!)

1 large dandelion leaf

2 lamb's quarter leaves

1 sprig purslane

AMARANTH WRAP

1 large amaranth or red root pigweed leaf

2 sprigs purslane

1 sprig chickweed

IMMUNE WRAP

1 medium sorrel leaf

2 peppermint leaves or small peppermint sprigs

3 small oregano leaves

3 lemon thyme leaves

2 comfrey buds or blossoms

FRUIT TART

1 fig

3 blueberries

1 sprig spearmint

Crack open the fig and squeeze the blueberries and spearmint sprig into the crack.

PEPPER BEAN HUMMUS WEDGES

Hummus

1 large bell pepper, cut into 5 long wedges

3 to 5 green soybeans per slice

Finely chopped cilantro

Smear the hummus on the pepper slices. Press the soybeans into the hummus and sprinkle with the cilantro.

AN IMPROMPTU HARVEST PARTY FOR FAMILY AND FRIENDS

My family and close friends love any excuse to get together for a good meal. Rather than plan an elaborate menu weeks in advance, we often head straight to the garden that morning, pick whatever is ripe, and start inviting guests while a menu based upon these brand-new garden gifts is percolating in our heads. We'll make a quick trip to the farmers' market or natural foods store to see what else looks fresh and good to supplement our backyard ingredients. Then we'll sketch out a quick plan, pull out our pots and pans, and start chopping, making things up as we go.

Often we expand the Farm Nibble concept by setting up a wrap bar, with a choice of fresh lettuce leaves and raw or lightly cooked collard leaves for the wrappers. We'll put out a cold buffet of chopped and sliced bits of raw vegetables, as well as organic cheeses, homemade hummus, and guacamole, to fill the lettuce. For the collard leaves, we will offer hot fillings such as bean stews, grilled meats or chicken, and grilled or roasted veggies. Crumbled feta or goat cheese makes a nice addition.

We find that even the youngest kids in the family, who typically are picky eaters, will go for one of these healthy treats. To round out the meal, my dad likes to make biscuits from scratch, and often guests will bring contributions of breads and desserts. If we are entertaining at my parents' house, my mom can quickly set a festive mood on the poolside deck by covering the patio tables with sheets of hemp and decorating them with wooden containers of herbs, colorful peppers, and candles. And often, if the spirit moves us, we'll pull out the many drums and other instruments scattered about the household and have a little family "jam" for dessert.

★

MASSAGED KALE SLAW

Of all the things we make, this is what we eat more than anything else. The kale is not cooked; instead, we soften the leaves by massaging them with salt, olive oil, and lemon juice. We dedicate a large part of our backyard garden to growing a few different varieties of kale, including lacinato, Siberian, and Red Russian. Once the kale is rubbed and seasoned, consider adding chopped tomatoes or avocadoes, soft- or hard-boiled eggs, or nutritional yeast.

MAKES 2 TO 4 SERVINGS

1 BUNCH KALE (LACINATO IS OUR FAVORITE BUT ALL VARIETIES WORK FINE), STEMS REMOVED

¼ CUP (60 ML) EXTRA VIRGIN OLIVE OIL

JUICE OF 1 LEMON

¼ TEASPOON SEA SALT, OR MORE TO TASTE

1. Remove stems and finely chop the kale leaves; place them in a bowl.

2. Add the olive oil, lemon juice, and ¼ teaspoon salt.

3. Toss and massage with your hands until the kale is soft. Taste, and add more salt if desired.

★

WHIPPOORWILL PEA AND BUTTERNUT SQUASH STEW

I was introduced to this delicious heirloom Southern pea variety by Andy Byrd, a friend whose farm goes by the name of Whippoorwill Hollow. One of my favorite ways to eat these Southern peas is in this delicious stew Stephanie concocted, which is also great as a filling for collard burritos.

MAKES 4 TO 6 SERVINGS

1 CUP (225 G) DRIED WHIPPOORWILL, BLACK-EYED, OR OTHER FIELD PEAS

2 TABLESPOONS OLIVE OIL

1 LARGE ONION, CHOPPED

2 TO 3 CLOVES GARLIC, MINCED

1 (2- TO 3-INCH/5- TO 7- CM) PIECE KOMBU (OPTIONAL)

5 SPRIGS FRESH LEMON THYME

1 SPRIG FRESH ROSEMARY

1 MEDIUM BUTTERNUT SQUASH, PEELED AND DICED

2 TABLESPOONS TAMARI OR MORE TO TASTE

1. Soak the peas, in enough water to cover them by 2 or 3 inches (5 to 7 cm), for 8 hours or overnight. Drain and rinse.

2. In a large stockpot, heat the olive oil over medium heat; add the onion and garlic and sauté until tender. Add the peas, 1 quart (480 ml) water, kombu, thyme, and rosemary. Bring the mixture to a boil, reduce to a simmer, and cook, partially covered, until the peas are tender, about 1 hour.

3. Add the squash and cook until it is fork-tender, 10 to 15 minutes.

4. Season with the tamari.

"The shared meal elevates eating from a mechanical process of fueling the body to a ritual of family and community, from the mere animal biology to an act of culture."

Michael Pollan, *In Defense of Food: An Eater's Manifesto*

"Generosity is giving more than you can, and pride is taking less than you need."

Khalil Gibran

8.

SHARING = GENEROSITY

Growing, harvesting, and sharing food are perfect catalysts for cultivating community. I strongly believe that through collective farming and gardening we can solve some of the greatest economic and social problems of our time. In the previous chapter, I encouraged you to reward yourself for your hard work by gathering friends and family to your table to celebrate the bounty. In this chapter, I want to inspire you to extend that generosity well beyond your immediate circle of friends and family.

According to the U.S. Department of Agriculture, one in every eight families experiences hunger or the risk of hunger. Ironically, many poor people also suffer from obesity and diet-related diseases, such as diabetes and high blood pressure, because they live in "food deserts"—places where fresh, healthy food is not readily available and jumbo bags of potato chips and sugary sodas bought in convenience stores often constitute a meal. Thankfully, if you grow food, you can help.

As the surge in home and community gardening has begun to restore small-scale food-growing knowledge that has skipped generations, gardeners are increasingly sharing their bounty, sometimes growing 100 percent for those in need. Food banks and pantries, which used to supply mostly canned goods, are expanding to accommodate more and more fresh food. National and regional food-bank organizations connect gardeners with places to donate within their communities that can handle the challenges of getting fresh food to those who need it most.

Plant a Row for the Hungry is a public service program run by the Garden Writers Association Foundation that encourages gardeners to grow extra food in their gardens for donation to local food banks. Since its inception in 1995, the program has helped facilitate the donation of more than 16 million pounds of fresh produce. Set aside a row in your garden and you too can help feed those in need.

Equally rewarding is the sharing of knowledge, encouragement, and time with others who can benefit from what you have to give, as many gardening activists have discovered by helping start or participating in children's gardens in places like schools, YMCAs, or boys' and girls' clubs.

In this chapter we will explore not only the various avenues available for generous giving, but also some of the logistics of spreading the bounty—as well as the wisdom.

FROM FARM TO CITY

One of the most formidable challenges of farming in Wisconsin was figuring out how to make money through the long, cold winters. Just before the first frost came in and turned everything from green to brown, we would glean the fields of every last fruit and leaf that was still edible. It is a Jewish tradition called *Pe'ah* to donate "the corners of the fields" to those in need and we would do just that. Each week at the farmers' market, a group would come around after we had packed up for the day and take our unsold produce as donations for the local food bank.

It began to occur to me that in order to make a bigger impact, I needed to do more work in urban environments, and I started to lean toward becoming an ambassador for social change through urban agriculture. I wondered how I could bring the farm to the city and start influencing the masses to think more critically about their food choices and how they affect the environment, the economy, and human health.

I volunteered to teach gardening at a number of schools and joined the board for the Children's Gardening Network in Madison. My interest in nonprofit social justice and urban farming work flourished, and soon I was guerilla-gardening all over downtown Madison. I worked with a group of youth detainees, taught gardening at a Jewish day school, and led a garden activity every week for toddlers at a Buddhist preschool.

BUILDING ELEVEN GARDENS IN ONE DAY

In 2009, Comcast (a cable TV and Internet company) contacted me with an intriguing proposition. They wanted to know if I would like to help them build eleven gardens for Boys and Girls Clubs, a Latin American association, and a nature center. I jumped at this seemingly brilliant way to take the message of healthy growing viral.

What I soon learned is that they wanted over a thousand Comcast employees to do the actual work, and have me coordinate it so that it would all happen in a few hours, for a volunteer day called Comcast Cares Day. This was going to be televised, so it had to go off without a single hitch. I will do just about anything to get gardens growing in needy communities, so I tackled the challenge by putting together a strategy, timeline, and budget that could be executed.

It took months of preparation. The garden sites were spread out over a 60-mile radius and each one had specific nuances regarding layout, terracing, and what to plant. I provided detailed written instructions for each garden installation and facilitated training sessions at my store for the Comcast volunteers who were taking the lead at each site. I showed them how to level, drill, and lay out the beds, and then how to fill them, add fertilizer, and plant from both seeds and plants. I gave each of them a list of tools they would need, a sketch of the garden layout, and directions on what to plant and how. We spent the three days leading up to the event in a frenzy of organization and intense physical activity. Thankfully, the event was a huge success and I slept for three days straight afterward!

I haven't attempted to build eleven gardens in a day again, nor do I plan to. But my company has partnered with Captain Planet Foundation and Whole Foods Market to build dozens of gardens at schools all around Atlanta. Whole Foods will often host a "5% Day" at one of their stores, where 5 percent of that day's net sales are donated to a local nonprofit or educational organization. This can raise up to $3,500 to build a school garden; as Whole Foods' partner, we provide the design, materials, construction, and programming to implement these projects. Our partnership with Whole Foods is extra special because the Farmer D Organics soil we use in the gardens is made in part from Whole Foods scraps, which is a great story to tell the kids.

Ever since we opened our flagship retail garden center in Atlanta we have extended discounts and donated products and services to schools and other nonprofit organizations that are trying to start or improve gardens. We have also helped a local homeless shelter in downtown Atlanta develop a rooftop garden to grow food and provide a healing space for the residents. We consider this to be an important part of what we do, and it has more than paid for itself by strengthening our relationship with the communities we serve.

★

TOOLS FOR SHARING

When there are lots of mouths to feed and it's time to share the garden's abundance, here are a few tools that can help:

SOUP POT: Every gardener and foodie should own a big soup pot so they can cook up the abundance of homegrown, farmers' market, and CSA veggies. Making soup is a great way to share the harvest with others, whether it's by inviting people over for dinner or sending them soup-filled Mason jars as gifts. At home, we use a 4-quart stainless-steel soup pot for just the family, an 8-quart if we are having a few guests over, and a 20-quart if we are cooking for the soup kitchen.

GRILL: There are few better ways to share the summer harvest than by cooking out on the grill. My favorite crops for grilling include eggplants, squash, peppers, tomatoes, onions, and corn. There are plenty of basic and affordable grill options at your local hardware store, but I suggest investing in one that does your homegrown organic veggies justice; I particularly recommend the Big Green Egg. These dome-shaped, ceramic-lined grills are modeled after the old clay cooking vessels used during the Qin Dynasty in China and by the Japanese as early as the third century. They are a little pricey, but it is hard to beat grilling with natural charcoal and slow cooking on cast iron or ceramic. While the Big Green Egg seems to be the leader in this type of grill, there are some similar options, such as the Primo Grill, Komodo Kamado, and Big Steel Keg.

ACCESSORIES: In order to host a large party and share the abundance with others, make sure you have a few large bowls, ample serving utensils, and plenty of dishes to go around.

YERBA MATE: A gourd (*guampa* or *cuia*) and accompanying straw (*bomba* or *bombilla*) for passing around yerba mate tea is a wonderful addition to any get-together, large or small. Mate is a vitamin- and mineral-rich tea that has significant healing properties, thanks in large part to its high levels of antioxidants. Add a few sprigs of fresh peppermint, lemon verbena, or lemon balm to brighten up your mate.

CONTAINERS: When sharing your harvest on the go, I strongly recommend using eco-friendly containers and utensils, such as To-Go Ware, to avoid adding plastic, Styrofoam, and other non-renewables to the landfill.

★

SUMMER VEGETABLE GRATIN

My sister Cindy is constantly experimenting with recipes to make use of her backyard abundance, and this is one of her favorites—to serve to her family or tote to a potluck. She got the idea from superstar chef Thomas Keller's cookbook Ad Hoc at Home. *Our version's simpler and quite a bit more rustic.*

MAKES 4 TO 6 SERVINGS

¾ CUP (180 ML) EXTRA VIRGIN OLIVE OIL, DIVIDED

2 LARGE YELLOW ONIONS, COARSELY CHOPPED

2 LARGE GARLIC CLOVES, MINCED

KOSHER SALT

2 TABLESPOONS MINCED FRESH THYME, MARJORAM, OR OREGANO LEAVES, OR A COMBINATION, DIVIDED

¾ CUP (75 G) FRESHLY GRATED PARMESAN CHEESE, DIVIDED

½ CUP DRY UNSEASONED BREADCRUMBS

1 MEDIUM YELLOW SQUASH, TRIMMED AND CUT IN ¼-INCH- (6-MM-) THICK ROUNDS

1 MEDIUM ZUCCHINI, TRIMMED AND CUT IN ¼-INCH- (6-MM-) THICK ROUNDS

1 JAPANESE EGGPLANT OR OTHER SMALL NARROW EGGPLANT, TRIMMED AND CUT IN ¼-INCH- (6-MM-) THICK ROUNDS

FRESHLY GROUND BLACK PEPPER

3 TO 4 MEDIUM-SIZE ROMA (PLUM) TOMATOES, TRIMMED AND CUT IN ¼-INCH- (6-MM-) THICK ROUNDS

1. Preheat the oven to 350°F (175°C).

2. Heat ¼ cup (60 ml) of the oil in a large skillet over medium heat. Reduce the heat to medium-low, add the onions and garlic, and season with salt. Cook without browning, stirring occasionally, until the onions are translucent, 15 to 20 minutes. Stir in 1 tablespoon of the herbs.

3. Set aside ¼ cup (25 g) grated Parmesan cheese. Combine remaining ½ cup (50 g) Parmesan with breadcrumbs and remaining tablespoon of minced herbs.

4. Spread the onion mixture in the bottom of a 13-by-9-by-2-inch (33-by-23-by-5-cm) baking pan or large round or oval gratin dish.

5. Arrange yellow squash, zucchini, and eggplant slices in a single layer in rows, slightly overlapping, to completely cover the onions. Drizzle with another ¼ cup (60 ml) of the oil. Season lightly with salt and pepper.

6. Sprinkle evenly with half of the herbed breadcrumb and cheese mixture. Spread tomato slices over the top. Drizzle with remaining oil. Season lightly again with salt and pepper.

7. Sprinkle with the remaining breadcrumb and cheese mixture.

8. Bake for 1¼ to 1½ hours, or until the vegetables are completely tender when pierced with a knife. Remove from the oven and allow the gratin to rest for 10 minutes. Turn on the broiler.

9. Just before serving, sprinkle gratin with reserved ¼ cup (25 g) Parmesan and place under the broiler for a minute or two until cheese is lightly browned. Serve warm or at room temperature.

★

SHARING MEDITATIONS

The obvious reward to growing one's own food is being able to eat it. Over the years, I have had the opportunity to learn how to cook and share my abundant harvest with others. It is in this preparing and sharing process that my gratitude truly comes to life. There are few things I get more excited about in life than bringing a harvest from the garden to the kitchen, and then watching as other people enjoy fresh, tasty food.

To me, every aspect of cooking is a meditation and a reminder to be thankful for the joy of sharing the bounty.

WASHING: As you wash your veggies, be thankful for the water that grew your harvest and now cleans it. Bathe your crop as you would a baby, gently scrubbing each and every crevice to ensure all the bugs and grit are removed. Take your time and embrace this moment as another opportunity to let your mind rest in gratitude.

CHOPPING: Use a clean, sharp knife and remain aware, to ensure all you chop is the crop, and not any part of your fingers. Focus on the rhythm of the blade against the board. Think about the positive energy you want to impart to this dish. Take this as yet another opportunity to rest, reflect, and be grateful—and careful.

COOKING: You have worked so hard to get the crop to this point, and now is the time to make sure it doesn't go to waste. Take care not to overseason, overcook, or burn. Once again, be mindful, be generous, and be present.

"ORDER OF SEEDS"

In a Jewish text about agriculture called *Seder Zeraim*, "Order of Seeds," six laws about gifts to the poor, are outlined. Here they are, with very brief descriptions:

1. *Pe'ah*, Hebrew for "corner," is an ancient Jewish law that says to leave "the corners of the fields" standing for the poor to harvest.

2. *Leket*, meaning "gleanings," are the ears of grain that fall from the harvester's hand or sickle and are gifted to the poor.

3. *Shich'chah* are "forgotten sheaves" that are left in the fields and gifted to the poor.

4. *Oleilot* are "immature grapes" that are gifted to the poor.

5. *Peret* are clusters of grapes that fall while being harvested and are left for the poor.

6. *Ma'asar ani* is the tithe designated for the poor every third and sixth year of the tithing cycle.

FIVE TIPS FOR GROWING FOR THOSE IN NEED

There are things you can do to make it easier for your donation to be used at its peak of freshness, without creating extra work for volunteers. Here are some tips:

1. FIND OUT THE SPECIFIC NEEDS OF THE POPULATION SERVED BY THE FOOD PANTRY TO WHICH YOU ARE PLANNING TO DONATE. Pay particular attention to culturally specific food preferences so that you are growing appropriate fruit and vegetable varieties. Consider what is not available, or not available affordably, in local markets.

2. GIVE MORE OF ONE THING RATHER THAN A LITTLE OF MANY THINGS. Everything you give will most likely be appreciated, but a handful of potatoes or a bag of cooking greens will feed a family more adequately than three radishes and a couple of carrots will. Go for quantity when you plant for those in need.

3. KEEP IT SIMPLE. Except for specifically desired or culturally appropriate crops, keep the varieties pretty darned basic. Growing for those in need is not the time for the fancy lettuces and lemon cucumbers, unless you intend to be there to explain what everything is and how to use it.

4. PACKAGE ITEMS APPROPRIATELY. Find out specifically how the food pantry distributes fresh food, and package your donations in a way that matches the system. Some pantries put fresh food in buckets or baskets on tables, and the food pantry clients choose for themselves. Others like to give out washed and weighed family-size bags. Ask if there is refrigerated storage, if food needs to be washed, when is the best time to drop things off, and if there are other practices that work well for the pantry to which you donate and the clientele that is served.

5. DO NOT DONATE ANYTHING THAT YOU AND YOUR FAMILY WOULD NOT USE. The food pantry is not the place for the rotting tomatoes and the tasteless, oversized zucchini! Of course, if it's something in good condition that you just don't like, then share with those who might.

6. GET INVOLVED. Volunteer to help with the actual distribution of the food. You will find out firsthand what works and what doesn't, and you may even be able to propose ways to do things more effectively. You most likely will discover that providing fresh food to those in need is more fun and uplifting than you think, and you will certainly learn more about both gardening and humanity (as well as how to say "lettuce" and "potato" in several other languages).

HOW TO PLANT GIFTS THAT KEEP ON GIVING

1. DONATE TIME. Consider hanging out at your local community or school garden. Ask what you can do to help, and just dig in, whether or not you are a member or parent. Offer to teach a class or serve as a mentor to a beginner. Don't think you know enough? Please realize that anything you know is more than many people know; some folks do not even know how to plant a seed. Your knowledge, no matter how vast or limited, is valuable.

2. PLANT A TREE. What more enduring symbol do we have than trees? Many urban areas have community volunteer groups that plant trees throughout the city, and many are specifically planting orchards. Plant a tree to provide sustenance for your children and grandchildren, while teaching them the virtue of generosity.

3. GIVE WHEN YOU'RE GONE. Established community gardens and nature centers in your area may offer something called "planned giving," which is a way for you to leave a contribution as part of your final wishes. Why not ensure that all of your hard work pays off in a way that gives back and keeps your money growing—literally—for many, many years? Plus, I can guarantee you that it will be appreciated, as most gardens operate on a true shoestring budget.

4. SHARE THROUGH SOCIAL MEDIA. Start a blog, share photos of your garden on social media, and connect with friends and family near and far in ways that old-time gardeners never could have imagined. By doing this, you plant a legacy of knowledge recorded for posterity.

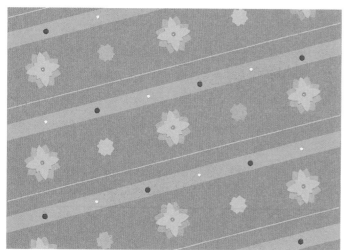

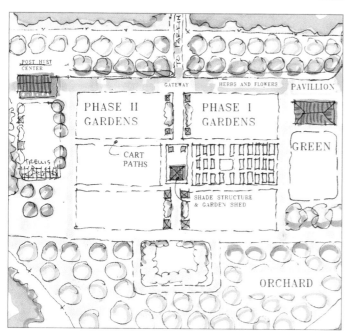

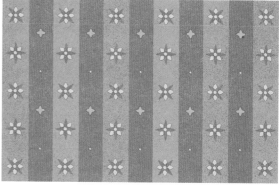

TAKE YOUR GARDEN VIRAL

I'll be the first to admit that I prefer the tweet of baby chicks to that of my Twitter account. That said, social networking is a wonderful way for gardeners and farmers to exchange ideas, images, projects, and tips with our followers and their friends. Many people and companies are using these platforms very effectively to promote themselves and their businesses. While that is a benefit, I believe the real value is being able to raise awareness, increase the number of conscious consumers and producers, and connect them with one another. As an example, we just completed a social media campaign to raise money for a homeless shelter and successfully reached our $2,500 goal to expand their rooftop garden.

To network with other citizen farmers, share stories and lessons, and work together to impact communities, go to FarmerD.com and join the Citizen Farmers Network.

If you are involved in a community or educational garden, the Internet matters, especially if you are trying to garner support, recruit volunteers, get business sponsorships, and find much-needed resources. Many of these gardens already have a website, but how many are truly putting the power of social media to work for them? A blog, Facebook page, Twitter account, YouTube channel, and Pinterest board are all ways to tell your garden's story, attract members, generate good press, encourage donations, and connect with others both close to home and around the world.

Proactively incorporating social media into your garden's communications efforts not only lets you manage your garden's reputation but can also deliver some effective outcomes, including:

SHARING KNOWLEDGE: Gardening knowledge has skipped not one but two (some would say three) generations, and many gardeners need a great deal of assistance to keep their gardens productive. Posting videos, articles, and other educational information is a convenient, accessible way to "handhold" new gardeners with busy schedules.

COMMUNICATING PLANS: Sharing dates, times, polls, agendas, and other member information via social media can be very convenient, as everyone can see comments and respond in real time.

CONNECTING WITH A LARGER COMMUNITY OF GARDENERS: Social media knows no boundaries and lets gardeners from many gardens share information and resources, learn from each other, volunteer at each other's gardens, and advocate en masse for needed agricultural ordinance changes.

RAISING MONEY: Social media offers a fast way to ask people to participate in online grant competitions or corporate sponsorship challenges, or to donate to a garden via secure online payment options. It also enables business contacts to see what some of the garden's needs are and to donate money or even volunteer labor.

VOLUNTEER: If you have extensive social media experience, you may be the perfect person to head up an effort at your community, school, or company garden. If you have little experience in this area but would like to gain more, you may find that volunteering as a social-media point person is a low-risk way to build some real business skills that may even help you get promoted at work.

HOW TO OVERCOME THREE COMMON SCHOOL GARDEN CHALLENGES

The school garden movement has surged in our country, and knowledge that has skipped generations is now being restored. This is very welcome news. What is a bit sobering, however, is when gardens that were started with good intentions and high hopes find themselves abandoned or poorly tended after just a few years. Sometimes this happens because the parents who served as champions have moved on as their children have grown up. Other times budget cuts may have limited the available resources for garden upkeep. Finally, pressure to "teach to the test" may keep children indoors working on classroom exercises rather than digging and planting. I'd like to offer some tips for overcoming each of these common challenges.

1. IT TAKES A VILLAGE. Let's face it—many times it takes a passionate parent on a wild tear to muster up the energy and support to make a school garden happen. This creates a need for continuity when that parent or team of parents is no longer available. Encourage a teacher to truly "take on" the garden. Designate a point-person parent in every class or grade. Encourage teaming across the educational continuum, where students of all ages help out in their next school or their previous school so that knowledge and support aren't lost with each new year.

Another great long-term solution to this common problem is to partner with a local garden center, or, even better, hire a gardener who will serve as a consistent source of expertise. Many schools are finding ways to afford a part-time or even full-time garden educator who cares for the garden and helps lead classes. There are also some companies, like my own, that offer ongoing garden services for schools, including everything from seasonal maintenance to training teachers and leading programs for kids.

2. MONEY DOES NOT GROW ON TREES. Budget cuts make garden resources hard to fund. This is an opportunity for one of the greatest life lessons of all: how to get creative. First of all, do the obvious things like applying for school garden grants. Some good resources include organizations like Kids Gardening, Captain Planet Foundation, and Whole Kids Foundation, as well as helpful websites like gardenabcs.com and edibleschoolyard.org.

It is a good idea to put together a fund-raising campaign, and don't overlook the word "fun" in that! I have found partnering with local restaurants and grocery stores to be a very effective way to raise funds fast. Restaurants will often be more than happy to host a fund-raiser dinner to support a school garden, especially if the owner's children go to that school.

Have a garden art sale, a "music in the garden" event, or a "dance in the dirt" party. Ask grandparents to give a "legacy gift" to your modern-day Victory Garden. Sell bricks with names on them for a Yellow Brick Road garden path. Sell garden books. Ask local businesses to hold "spirit days" at their stores, where a percentage of sales goes back to the garden fund. Have a dress-down or funky-hat day and charge a dollar or two, with proceeds going to the garden. These are just a few examples to get your juices flowing. Remember to ask the students for their ideas, too—they are often your best advisors!

3. "GO TO THE GARDEN!" Wouldn't it be great if when you did something good in class you were sent to the garden? A nice counterpoint to being sent to the principal's office when you do something bad! Once a school has invested in a garden and made the effort to keep it growing, it is a shame if it doesn't get used year after year.

Most teachers are required to follow state standards and think that they don't have time to "waste" in the garden. On the contrary, integrating the garden into the school curriculum will provide an abundance of opportunities to teach valuable lessons in many subjects, from math and science to social studies and language arts. Hands-on learning in gardens has been shown to encourage children of all ages to explore, experiment, and experience; to make valuable connections across the curriculum; and to learn in the way that best fits their individual needs (including a wide range of special needs). A class parent can help find specific lessons plans to help the teacher be especially efficient with his or her gardening time and ensure that the students "stay on task."

No matter what challenge your school garden is up against, there is certainly some other school that has faced—and overcome—that challenge. Let's learn from each other, and help our children continue to grow in a broad spectrum of ways.

TIPS FOR WORKING WITH CHILDREN IN SCHOOL GARDENS

You may have been excited to work with a group of children in a garden, only to discover that the experience did not live up to expectations, for you or for the children. You may have overestimated what you could accomplish, or the children may have found your structured lesson plan too structured. You may have planned very little and then realized that the children could have used more guidance. You may have miscalculated the stage of development of the age group with whom you were working, or you may simply have needed more adult help or more tools and supplies. Don't give up. Learn as you grow, so to speak, and try again. Here are some tips that may help:

1. REMEMBER THAT MANY CHILDREN MAY NEVER HAVE BEEN IN A GARDEN. This means they may not know basic garden etiquette, such as not stepping on (or running through) garden beds. They may be scared of some things, like bugs or bees. They may not be used to patiently observing things. And they need your help to learn about all these aspects of a garden and how to behave safely and respectfully. Keep it clear and simple and kids tend to get it.

2. YOUNG CHILDREN ARE HAPPY TO PUTTER, AND THEY HAVE SHORT ATTENTION SPANS.
Don't get hung up on long lesson plans for the little ones. They will be happy to dig a hole, plant a seed, paint a decorative rock, or try to catch a butterfly. Be aware that the mood you create with them in the garden could set the tone for how they feel about gardening for the rest of their lives. Make it fun. Make it memorable and let them enjoy discovering nature.

3. FOLLOW THE CHILDREN'S LEAD TO SEE WHAT THEY ARE INTERESTED IN. If they find an inchworm, use that as an opportunity to teach about measurement. If they seem fascinated with how fast a bean seed germinated, that's a great time to talk about the parts of a plant and how seeds grow. Toss in a little history nugget here and there, such as, "Did you know that people carried seeds with them when they moved to a new country?" Show math in nature: "Look at how the beehive is made up of hexagons!" Create art in the garden: "Let's see how many different colored dyes we can make from plants." Tell stories. Sing songs. Dance. Pretend. Senses are heightened out in nature, and the children will learn more than you can imagine.

4. MIDDLE SCHOOLERS NEED TO FEEL NECESSARY. They are at an age when they are not quite sure where they fit in. This is also a time when they start to feel a sense of environmental responsibility. This coincides with them being at an age where they are strong enough to do real work. Take advantage of these crossroads in their lives to focus their attention on solving real problems and achieving something tangible.

They are ready now to be presented with goals such as "Create a new twenty-five-foot row," or "Find ways to save more rainwater," and to figure out for themselves how to achieve them. Give them the opportunity to create their own vision so they truly feel ownership of it. Jump in to help where needed, but don't solve the problems for them (and do let them make mistakes). You'll see their self-esteem, decision-making capabilities, and teamwork skills soar.

5. HIGH SCHOOLERS ARE READY FOR SERIOUS PROJECTS AND LEADERSHIP. Students of this age are ready to tackle higher-level academic subjects in depth and to propose innovative experiments that test their knowledge and explore what's possible in a changing world. A high school garden program is a terrific place for a STEAM emphasis (science, technology, engineering, art, and mathematics), as students computerize irrigation, build solar systems, propose biomimicry inventions, and study all aspects of environmental science. Artists can find inspiration in the garden, journalists can cover the garden story through multimedia, film students can document it, language students can build communications connections through signage and multicultural events, math students can create budgets and planting plans, and those with special needs can benefit from sensory integration in the garden to create pathways for future success. In fact, I can't think of anything that can't be taught in some way in a school garden! If you listen, truly listen, to what children—of all ages—tell you about their interests, the garden can help them grow in those directions.

★ TOP 10 ★

MOST ABUNDANT CROPS FOR SHARING

Looking to make friends? Here are some great crops that can produce enough to share with others and provide plenty of food to go around:

1. **SUMMER SQUASH:** Zucchini, yellow crookneck, and pattypan are incredibly prolific and will likely give you more fruit than you can keep up with. I don't know many good ways to preserve summer squash other than zucchini bread and dried chips, so this is a great crop for feeding your friends, neighbors, and those in need.

2. **WINTER SQUASH:** This is an abundant crop with excellent storage life that is also a powerhouse of vitamins A and C and other nutrients, making it a great option for the food bank garden. Most winter squash need some room to sprawl, as the fruits set on vines—with the exception of acorn squash, which is a bush similar to a summer squash plant, making it a good choice for a space-limited urban garden. The vining squash and pumpkins are great to grow down slopes or over fences. Blue Hubbard is a variety that produces huge fruits, ideal for sharing.

3. **CUCUMBERS:** These prolific producers are pretty easy to grow and certainly delicious to eat. For the donation garden, I recommend growing the slicing varieties like Marketmore—save the pickling and lemon cucumbers for the kitchen garden.

4. **TOMATOES:** Not the easiest to grow, but excellent producers of yummy and nutrient-dense food, and when things go well there is often a bumper crop. While I am a huge fan of all the amazing heirloom tomato varieties out there, I recommend growing a few disease-resistant hybrids to ensure a steady harvest for your food pantry garden. Some reliable varieties include Rutgers, Celebrity, Big Beef, Juliet, First Lady, and Goliath.

5. **LETTUCE:** When lettuce season comes around in spring and fall, plant a bumper crop of romaine lettuce and share the love with others. Harvest large outer leaves or whole heads before they bolt.

6. **SWEET POTATOES:** One of my all-time favorite crops to grow. It always amazes me how a withering "slip" can take root and, even with minimal care, grow pounds and pounds of nutritious sweet potatoes. They do well on relatively poor ground and produce significant quantities of tasty food that stores well, making them perfect for the donation garden. A good variety for the backyard is the Bush Porto Rico, as it is more compact than the typical vining varieties like Centennial and Georgia Jet.

7. **LEAFY GREENS:** There are many vitamin- and mineral-packed greens that are easy to grow and great for sharing with others. The most abundant variety in my backyard right now is Siberian kale, but we also love to grow collards, mustard greens, and turnip greens.

8. **TURNIPS:** These nutritious roots do well planted amid a cool-season cover crop like vetch, clover, peas, or rye. Sow in late fall and harvest roots throughout the winter and into spring. If you grow in rows, plant about 8 seeds per foot in rows 12 to 18 inches apart. My favorite variety to grow on a large scale like this is the Purple Top White Globe Turnip, which can reach maturity in just 55 days.

9. **POTATOES:** There are few crops that can fill as many stomachs as good old-fashioned Irish potatoes. Everyone loves a warm baked or mashed potato.

10. **CARROTS:** These popular roots can be direct seeded close together in the garden and will complement your tomato and lettuce crops. They also have excellent storage life and nutritional content, making them a good crop for the sharing garden. My favorite varieties for high yields include Danvers 126, Nectar, and Bolero.

★ TOP 10 ★

BEST CROPS FOR KIDS

*If you want to get your kids excited about eating fresh vegetables,
I recommend starting with these easy-to-grow and yummy-to-eat favorites:*

1. CHERRY TOMATOES: These tasty little bursts of sunshine are perfect for school and home gardens, as they produce a prolific quantity of bite-size 'maters. They are the sweetest of all tomatoes and can be a gateway to the meatier varieties like slicers and beefsteaks. My favorites are Sungold, Black Cherry, and Matt's Wild.

2. CUCUMBERS: Cukes are prolific producers and are refreshing and healthy to eat. Slices of fresh-picked cucumber are usually a big hit with kids. I recommend letting them taste cucumbers with and without the peel, as they are less bitter when peeled. Some favorite varieties for a kids' garden include lemon, Marketmore, and Armenian cucumbers.

3. FRENCH SORREL: Known in many school gardens as lemon leaves for their tart, sour flavor, these hardy perennials are a must-have for school gardens and can be planted from seeds or plants. Prune back seed heads to prevent bolting and keep plants clean by harvesting often.

4. SUNFLOWERS: These majestic flowers add great character to a school garden, and their stems can also provide a sturdy trellis for other crops like pole beans, cucumbers, snap peas, and nasturtiums. Grow edible varieties and kids will have fun plucking out and eating the seeds. Some seed companies offer a Sunflower Snack Mix with a variety of the best edible sunflower seeds.

5. SNAP PEAS: Most kids love fresh peas and these are great for growing in a spring or fall garden, which is when most kids are in school. It's a great crop for them to plant from seed and watch grow to fruition. Some varieties like Cascadia, Oregon Giant, and Sugar Snaps can grow from seed to harvest in 55 to 70 days.

6. SNAP BEANS: This is another great crop for kids to start from seed and watch grow to harvest. It is important for children to taste the difference between a fresh-picked snap bean and the canned green beans they get in their school lunch. They are easy to grow directly from seed in the garden and can go a long way when chopped into small tasters for kids. For a nice colorful array, I recommend growing a mix of purple, yellow, and green snap beans.

7. VIOLAS: One of my favorite little school-garden snacks is a sorrel leaf taco with green bean and viola flowers. Violas are a nice way to add color to a school garden, and with their abundance of blossoms, they provide plenty of edible flowers to go around—for those brave enough to eat flowers!

8. CARROTS: While they do require a bit of patience, carrots are a delicious and fun crop to grow in any school garden. Most kids like carrots and identify them as both healthy and delicious. They also make a great backdrop for the Peter Rabbit stories! There are some larger, more unusual varieties that are great for the fall and winter garden, including the rainbow mix of red, purple, yellow, white, and orange carrots. Thumbelina is a quicker option that is fun for kids because of its midget size and round shape.

9. POTATOES: These little jewels of the earth are fun to plant and even more fun to harvest. I love watching kids dig in the dirt for potatoes—and they all love French fries, so it's easy for them to get excited about it! The one drawback to potatoes in a school garden is that they really need to be cooked before they are eaten.

10. BROCCOLI: While not the easiest crop to grow in a school garden, this is one that has a high reward. I hated broccoli when I was a kid, until I ate my first fresh-picked broccoli. Ever since then it has been in the top three of my favorite veggies. I recommend growing the De Cicco variety, as it continues to send out side shoots after the main head has been harvested. Broccoli does best in the cool weather of early spring and late fall, which works well for the school growing season.

> *"The care of the earth is our most ancient and most worthy and, after all, our most pleasing responsibility. To cherish what remains of it, and to foster its renewal, is our only legitimate hope."*
>
> Wendell Berry, *The Art of the Commonplace: The Agrarian Essays*

9.
SUSTAINING = PERSEVERANCE

After the harvest is over, the cycle begins again, and ideally our farms and gardens become more fertile, abundant, and profitable each year. Unfortunately, that doesn't always happen. The reasons are endless: pests, weather, critters, drought, flood, fatigue, financial strain, lack of adequate help . . . the list goes on.

Farming is fickle, like any business. Some years are better than others; it requires a great deal of hard work and perseverance to weather the ups and downs. But don't give up. The stakes are too high and the rewards too great. Focus on the soil and the experience, and you will grow in many more ways than you could imagine. In my opinion, long-range success is measured by our impact on society and the earth, creating meaningful jobs, and being able to take pride in what we do. It is truly an honor to do something you believe in that can provide sustained health, happiness, and fulfillment.

So how do we keep our gardens thriving over the long haul? The answer, I believe, lies in empowering our fellow citizens to support and preserve artisanal agriculture, making it the way of the future rather than a quaint relic of the past. Engaging individuals, institutions, businesses, and government is critical to ensuring the long-term viability of both urban and rural agriculture.

I'm sure you understand by now what I mean when I say that farming is about so much more than just cultivating food. It's about the relationship and understanding you build in the process—with the land, your inner self, and your community. If you practice the steps in this book, both you and your farm or garden will succeed over time, and you will be making valuable long-term contributions to the world around you.

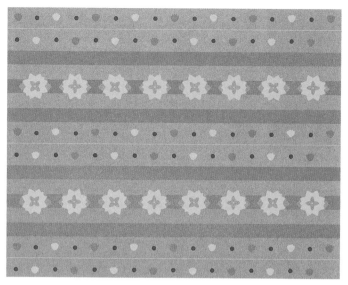

THINGS TO DO
IN THE GARDEN
Sing

In this chapter, I will talk about sustaining your projects and explore some of the challenges you will most likely encounter, and hopefully provide tips on how to persevere through them. I will also share ways to preserve the harvest and offer ideas for helping your gardening and farming endeavors to succeed into the future, whether you are seeking primarily to feed your family from your own backyard, to add to your income, or to contribute to the larger mission of supporting the sustainable agriculture movement.

FARMER D'S TIPS FOR THE ENTRE-MANURIAL SPIRIT

For those of you wanting to turn your gardening passions into a source of revenue, here are a few quick ideas to channel your inner entre-manure. Whatever you choose to do to cultivate a better world, I hope you do it with a smile and share it with as many people as you can. Together we can sow seeds for a brighter future and inspire coming generations of citizen farmers to do the same.

COMPOSTING: There is nothing more valuable in a garden than compost. Many people are not able to produce as much as they need at home or need a place to send the organics they generate. This creates an opportunity for the entre-manure to step in and help them. One way is to offer a compost collection service, where, for a small fee, you come to someone's home once or twice a week and pick up their organics. There are a number of models like this that have popped up all over the country. Whether you are looking for such a service or want to start your own, there's a black-gold mine here.

Worm farming is another business opportunity that often starts out with a bin or two in the basement and can quickly escalate to renting a warehouse for millions of worms. There is a market for worms for composting and fishing and an even larger market of gardeners who know the value of high-quality vermicompost.

Lastly, making quality compost on a large scale and selling it to gardeners, landscapers, and farmers can channel your love for compost into a viable career opportunity. There are some innovative composting companies that are making a big impact—and big profit—by redirecting organics from the landfill to their compost facilities and ultimately to a gardener like you.

PLANNING: Is there a designer in you? There are millions of homes and businesses out there that want to create new sustainable farms and gardens or improve their existing ones. Here is where you come in. The growing interest in creating edible, organic, and permaculture landscapes is providing opportunities for those of us who have knowledge to share and can help illustrate the possibilities. You don't have to be an artist or landscape architect to do this. You do need to have a good idea of what plants will do well in a particular area and be able to translate that into a functional and aesthetically pleasing design. Expressing a garden design on paper or through a digital program is a very valuable skill and one people are willing to pay for.

TILLING: More and more people are thinking about turning their large lawns or small backyards into food-producing gardens but are not necessarily up for the hard work that this entails. They are, however, willing to pay someone else to do the heavy lifting so they can start planting their way to food independence. Do you have a strong back and some business savvy? Maybe you made a few bucks mowing lawns in high school and are ready to take it to the next level. Invest in a few tools, a rototiller, and maybe even a tractor and trailer. You could be in the garden installation business in no time. There's nothing more fulfilling than coming home from a hard day's work with mud-covered knees and the deep satisfaction of having added another garden to the world.

SOWING: The easiest way to succeed in gardening, especially for beginners, is by starting with a seedling rather than a seed. A seedling is a plant that has been grown in a greenhouse for a month or two and is ready to be transplanted into the garden; it will grow to maturity in a much shorter time than if it had been planted directly from seed. This gives the gardener a jump start and also sidesteps the most fragile time in the life of a plant, from germination to young adulthood. These hardy little "teenagers" also allow a gardener with limited space to maximize production by reducing the time each crop takes from planting to harvest.

"Sow," herein lies the business opportunity—growing edible "starts" for home gardeners and farmers. Yes, even farmers experience the same benefits, and many are buying "plugs" for crops that are harder to grow from seed and to get a head start on their season. Plugs are basically very young seedlings, usually about one month old and just a few inches tall when they are transplanted. Starts, or transplants, are about two months old and are in bigger pots with larger root systems.

Once you master the art of growing from seed in a greenhouse, in a hoop house, or even indoors, you have a valuable product that lots of gardeners are willing to pay for. In our retail store, every year we sell thousands of organic vegetable, herb, and edible flower transplants to home, school, restaurant, and community gardeners. One good way to get started with this is to grow enough starts for your needs, plant extra as a buffer, and sell any surplus to friends and family—or use them as barter.

GROWING: Many people love the idea of growing, build a garden, then realize they are either too busy or too inexperienced to give the garden the attention it needs to thrive. That's where a passionate, knowledgeable gardener like you comes to the rescue, by offering a maintenance program in which you tend someone's garden once a week and leave a basket of fresh-picked veggies on their doorstep. There are more and more models like this sprouting up around the country. In fact, Farmer D Organics is one; we maintain dozens of gardens for individuals, schools, hospitals, and restaurants around Atlanta. We also help some of them maintain their compost and worm bins, backyard chickens, and bees.

HEALING: Gardens can actually provide therapy—and not just from working in the garden and eating fresh veggies. Growing medicinal plants and making teas, tinctures, and more can actually be a way to support your garden hobby and maybe even turn it into a lucrative business. One valerian plant, for example, can easily yield about 20 ounces of tincture, which could sell for a total of $140. You can see how a small medicine garden could pay for itself in no time. If this path is of interest to you, I recommend taking a course and doing some research, as these plants are very potent and should be used properly. There are also a number of herbal products like salves, lip balms, and lotions that are not ingested so do not pose as much of a safety concern. Who knows? You could become the next Burt's Bees. Another healing route is a horticulture therapy degree or certificate where you could make a meaningful living helping people heal through gardening.

REAPING: Perhaps you want to farm as well as reap the lifestyle benefits that come with it: more fresh food than you can eat, a community of people to share it with, meaningful work, a farmer's tan—and no gym membership! If you have tried your hand at gardening and found not only that are you good at it but that you can't imagine doing anything else, I suggest taking a leap of faith and starting your own CSA (Community Supported Agriculture) farm or garden. The beauty of a CSA is that others who are confident in your ability to grow their food will invest in you, sharing the risks as well as the rewards. You can start with a small group of friends and family working a big backyard garden, and possibly advance to a leased piece of land or eventually even buy your own farm. I can think of no better career path than being a CSA farmer, but it's not a get-rich-quick scheme. In fact, the riches are the lifestyle more than the money, so do not go down this road if your agenda is making lots of money with minimal effort.

A CSA is a great foundation for a career in farming and can set the stage for many, if not all, of the other opportunities I have outlined in this section.

SHARING: A growing number of social entrepreneurs are finding creative ways to bring fresh produce into the food deserts and to use gardening as a tool for educating, empowering, and employing people in need. There are some great nonprofit organizations working to share sustainable agriculture practices and resources, anywhere from their own neighborhoods to the most remote reaches of the planet. If this is your calling, I recommend working for one of these organizations before you start your own. As a past founder and executive director of a nonprofit, I can assure you that, as in a garden, there will be plenty of planning, sowing, and growing to do.

SUSTAINING: Most of what you grow is perishable and needs to be eaten or sold very quickly. Some options for selling your fresh produce include farmers' markets, CSAs, restaurants, coops, or grocery stores. However, there is a way to add value to and extend the shelf life of your harvest: by making and selling value-added products such as sauces and salad dressings. You may start out in your home kitchen, making some pesto with your bumper crop of basil and garlic and giving it as holiday gifts to friends and family. Then, once you feel you have a marketable product, it's time to find a certified kitchen. Nowadays there are food hubs with shared commercial kitchens, or you can find a restaurant, catering company, or underused community kitchen at a church that is willing to rent space at a discount.

In this model you don't have to keep growing everything yourself; you can also support local farmers by buying their fresh produce and turning it into something pickled, fermented, canned, dehydrated, or preserved. There are many multimillion-dollar value-added food businesses that started out in a home kitchen or hippie commune—take Stonyfield, Annie's, and Ben & Jerry's as your inspiration, start whipping up a batch of your grandma's favorite recipe, and see where it leads.

TOOLS NEED SUSTAINING, TOO

I'll admit I've dug with my hands or grabbed a stick to make a planting hole when I didn't have a trowel handy, so, sure, you can certainly make do in a pinch when you don't have the right gardening tools. However, I'd be one unhappy gardener if you asked me to go for long without my good-quality trowel, stirrup hoe, or digging fork, as they make garden work far easier and more effective. Hang around seasoned gardeners enough and you will discover that they all have favorite tools they have cherished—and cared for—for years. Most gardeners eventually discover that cheap tools aren't worth it, since they bend and break and then have to be replaced. They've also learned that taking a little time to care properly for their high-quality tools makes sense; this way they don't rust or splinter, so they stay sharp and useful for years.

I recommend dedicating an hour or two to setting up a tool-care system that helps keep things in ready-to-dig shape, and helps keep you more mindful of your tool care on an ongoing basis. Here are some helpful hints:

1. ORGANIZE. Have a place for everything and everything in its place. This seems obvious, of course, but how many of us have tools here, there, and everywhere? Take a little time to clean out a spot and create order for your tools. Hang them on a wall in your garage or in a shed. Stack them on a large shelf. Put tall tools in a big bucket, handle side down, and small tools in a smaller bucket.

2. DRY AND OIL. After you are done using a tool for the day, give it a good cleaning and wipe it off with a rag dabbed with a little cooking oil. Doing this will help keep diseases, fungi, weed seeds, and insect eggs from getting spread around your garden. I find the easiest way to clean tools is by first blasting the soil off with a hose at maximum pressure. If you have some sticky clay on there, a bristle brush may be needed; otherwise, just wipe it down with a cotton rag. Some people store hand tools in a five-gallon bucket of coarse sand with a little oil in it. This will help keep metal parts from rusting.

3. LUBRICATE AND SHARPEN. At least once a year, rub all wooden handles with linseed oil (it'll soak right in), lubricate moving parts of pruners and wheelbarrows, and sharpen blades with something called a mill file. Wearing goggles is a good idea anytime you sharpen metal, by the way. It's also wise to secure tools in a vise on your workbench, if you have one.

4. PLAY IT SAFE. Any time you put tools with tines on the ground, be sure that they are facing down. This is especially true for shovels, rakes, and hoes. This is so nobody gets hurt by stepping on a tool and having it snap up and whack them. In general, try not to get in the habit (or break it if you already have one) of leaving tools lying around, even when you are using them. Stand them up in the soil or lean them against something a little bit out of the way. Be especially vigilant about safety when children are helping or nearby. Also teach children to keep tools close to the ground at all times, never above the knees, to avoid any head or upper-body injuries.

It is very common for dedicated gardeners to collect quality tools over many years. Tools make excellent gifts that give back in garden bounty, season after season.

SHUCKING MEDITATION

Shucking and shelling bring back memories of sitting on the front porch of the farmhouse with harvest bins full of garlic or flint corn, or out in the grass with crates of leeks or green onions, relaxing after a long day of harvesting and just peeling away.

Sit down on a rocking chair with bins of garlic, leeks, onions, or corn. One by one, peel the layers and remove the outer husk and reveal the shiny crop below. You can sit and shuck for hours on end, letting the time go by while still being very productive. Shucking is also a great community activity that can be enjoyed with others.

★

TOOLS FOR PRESERVING

Now that your garden is pumping out more food than you can eat, it's time to start putting it up for the winter. Here are a few basic tools and you will need to can and preserve your way to total food independence.

CANNING POT: There are many options for canning pots, including a water-bath canner, a 16-quart stainless-steel pot, a pressure canner, a steam canner, or an electric water-bath canner, to name just a few. These canning pots range in cost from $40 on the low end to $400 for a fancy pressure cooker. I recommend starting with a 16-quart stainless-steel pot, as it can be used for other things like cheese-making and will cost under $100.

CANNING RACK: This stainless-steel basket fits into your canning pot to hold the jars upright and will cost about $25.

JAR LIFTER: This is essential for moving jars in and out of the boiling-hot water bath. It is easy to find and only costs about $6 for a basic model or $15 for a deluxe spring-loaded version.

CANNING FUNNEL: A canning funnel is a very useful tool when pouring hot brine or stewed veggies into pickling jars. The cost for a funnel is between $3 and $15.

CANNING JARS: You will need canning jars to put all your goodies in and there are a few types to consider. Ball (or Mason) jars are the most common, and they work great. I prefer the wide-mouth jars and like to keep a mix of quart and pint sizes. If you are making jams or jellies, you may want to stock up on a mix of 4-, 8-, and 12-ounce jelly jars.

ACCESSORIES: Some other little things that are good to have include a canning ladle, a magnetic lid lifter, a long spatula, and a canning jar brush. All can be purchased for a total of about $25.

FERMENTING CROCK: Fermenting is another great way to preserve your veggies and make healthy foods like sauerkraut and kim chi. Any lead-free glazed ceramic crock will work, or if you want to take things to the next level, consider investing in a Harsch Gairtopf fermentation crock. For home use, I recommend the 5-liter size, which costs about $100.

DEHYDRATOR: One of the best ways to preserve your harvest is with a good dehydrator. You can use it to dry fruit, veggies, and herbs and even to make homemade fruit leather. We love our Excalibur brand and use it often, especially during the harvest season, when the garden is kicking out more than we can eat. One of these can be purchased for about $250.

★ TOP 10 ★

CROPS TO PRESERVE

Grow for more than just the moment; think about the future, and prepare for it.
There are many ways to preserve the harvest—by canning, dehydrating, fermenting, or freezing.
This is a rewarding process and can provide wonderful gifts for your family and friends.
Here are a few of my favorite backyard crops to preserve:

1. **CUCUMBERS:** Make sure to grow a pickling variety and harvest often for tender, juicy pickles. I like to grow dill in with my cukes by direct seeding it down the center of the bed, then seeding or transplanting pickling cucumbers on each side of the dill. Some of the best pickling varieties include Northern Pickling, Salt and Pepper, Pickle Bush, and Regal.

2. **CARROTS:** Any carrots will work for pickles, but I like a mix of heirloom varieties for a nice blend of color and flavor. It's best to harvest carrots young so they are tender when pickled.

3. **BEETS:** The nutritious beet root makes an excellent pickle. First boil the fresh-picked beets for 1½ hours and rub off the skins. Some good varieties include Detroit Dark Red and Touchstone Gold.

4. **BUSH BEANS:** I love me some good old-fashioned dilly beans. Pickle your garden-fresh green beans and yellow wax beans with a few sprigs of dill and plenty of garlic and some hot peppers. You can also can fresh-picked beans for winter storage, which will make you hesitate to buy store-bought beans ever again. Tender-ette is a great bush variety and I like Blue Lake for a pole bean.

5. **CABBAGE:** A fall crop of late-season cabbages will make a delicious fermented sauerkraut with all kinds of health benefits. Some good varieties for kraut include Golden Acre, Kaitlen, Early Jersey Wakefield, and Late Flat Dutch.

6. **PEPPERS:** There are lots of amazing varieties of sweet and hot peppers to grow, with which you can make sauces of widely varying flavors and Scoville heat units (which measure the content of pungency-producing capsaicin). Early jalapeño, habanero, chili, and cayenne peppers are some of the most popular for making hot sauce. Some great pickling varieties include Hungarian, Cubanelle, Sweet Banana, and Sweet Bells.

7. **TOMATOES:** These have all kinds of characteristics that can make a tasty sauce, salsa, or sun-dried tomato. Some classic varieties for drying and sauces include Roma, San Marzano, Principe Borghese, and Napoli. Other varieties especially good for canning are Bellstar, Jubilee, Rutgers, and Tropic.

8. **OKRA:** Grow in warm climates through the summer and harvest often to keep production going. Once you are harvesting every few days, you will see the need for pickling. My favorite varieties to pickle are Hill Country Red and Clemson Spineless.

9. **BASIL:** To keep basil growing and not bolting, it is critical to harvest often, and a great way to preserve it is by making pesto and freezing it. The Genovese varieties are the most popular for pesto. Pesto is a great way to reuse the last of the basil crop before the first frost.

10. **BERRY AND FRUIT TREES:** Strawberries, blueberries, blackberries, raspberries, and figs are some good examples of homegrown perennials that are perfect for putting up in jams and preserves.

THE FALL GARDEN

If you work or travel a lot over the summer, your garden may seem overgrown and neglected, and may therefore be the last place you want to spend time. Follow these easy tips and I'll help you whip it back into shape so you can "fall" in love with your garden again:

1. CLEAN OUT PLANTS THAT HAVE HAD IT. There is only so much you can ask of a plant, and if it is showing signs of distress and you've already tried what you could to help it, just move on. It's time. The good news? Pulling up the big prima donna plants of summer is easy and clears a large amount of space very quickly. Also, the ground beneath the plants is usually not weedy because it has been thoroughly covered by your plants, so there is a strong chance you will see a satisfying difference very quickly. As for plants you think still have a little fight left in them, prune their dead leaves, add additional supports where needed, toss some compost and organic fertilizer around the base of the plant, and use their towering height as an aid and shade for your fall garden.

2. "PREP" YOUR BEDS FOR FALL. There are many ways to prepare your beds for the coming season. You could simply freshen up your soil with a top-dressing of new compost and organic fertilizer, and you're ready to plant. You could throw in a late-summer or fall cover crop to scavenge for nutrients and create additional organic matter to add structure and nourishment to your soil. If you plant a fast-growing late-summer cover crop like buckwheat, you would cut it down in about a month and let it decay for a week or two before planting your fall crops. If you plant an overwintering cover crop like Austrian winter peas, vetch, rye, and/or clover, you can let it go until early spring, then turn it under and get ready to plant a spring crop two or three weeks later.

3. TAKE A SOIL TEST. Dig up a soil sample and send it to your county extension office or other soil lab and ask for organic recommendations. You can get a report back in a week or two that will tell you what deficiencies your soil has so you can amend it with plenty of time before spring planting.

4. PLANT FALL CROPS. To find out what to plant when, look for a local planting calendar, which is often available at no cost through a university extension, garden center, or local organic farming nonprofit organization. An additional indicator of planting-readiness is soil temperature. Seeds all have their favorite soil temperature for maximum germination. Remember those hardy summer plants that you left in the garden in tip #1? The soil in their shade may be cool enough to plant fall seeds several weeks earlier than expected, and then they can provide nice shade cover for fall crops while the sun is still scorching the garden elsewhere.

EXTENDING YOUR GROWING SEASON

Many folks think once football season begins and swimming pools close, the vegetable garden is also shut down for the winter. This could not be further from the truth. Warmer areas of the country are fortunate enough to have an extended growing season due to (mostly) mild and short winters. Collard greens, broccoli, cabbage, parsley, onions, garlic, mustards, root vegetables, and many more cool-season crops are cold hardy and frost tolerant. In fact, the leafy greens get sweeter as the air chills, and the bugs just about disappear.

If you are not ready to surrender your basil and tomatoes to the chill of winter's inevitable arrival, bring out the frost blankets, row covers, or old sheets. These will allow you to fend off winter for a few extra weeks or even months.

Row cover keeps heat from the bed trapped in at night, raises the soil temperature quickly during the day, and also helps protect plants from the desiccating effects of strong winds. Using wire hoops or your own home-made supports to build a tunnel will help to hold up a heavier frost blanket over your raised garden beds, giving your plants the ability to grow upright unhindered. Floating row cover is an extremely lightweight, breathable fabric that can be laid directly over plants without a supporting framework, but it will need to be secured with garden staples or stakes to keep it in place. With a little care, these fabrics can be used for several seasons.

For a more permanent season-extending garden fixture, try using a cold frame, which is essentially a raised bed box with a slanted panel of glass or plexiglass on top. This "window" is on hinges and can be opened and closed easily. Cold frames are like mini greenhouses, allowing sunlight through but preventing heat from escaping. They are best located facing south, to capture the most light and heat. The valuable space within cold frames can be used in spring to germinate seeds or "harden off" seedlings, easing their transition from being in the comforts of a greenhouse to braving the elements of the unprotected outdoors.

By following the recommendations above, you can rest assured that as you are warming your toes by the fire and sipping homemade soup, your plants will be just as cozy as you are.

REPAIR THE WORLD—ALL WINTER LONG

What if I told you that you can help repair the world all winter long by barely lifting a finger? And what if I told you that this would set the stage (or, rather, prepare the ground) for extraordinary bounty in the year to come? Would that be worth an hour or two now, especially if those couple of hours were spent outdoors in the beautiful fall weather? Here are three things you can do this winter to prepare yourself and your garden for spring:

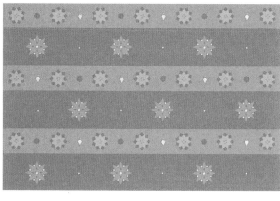

1. RECYCLE LEAVES. Save, crush, and spread dead leaves on your garden beds, and add some to your compost pile as well.

2. SPREAD THE LOVE. Give from your garden and from your heart. Share your knowledge and your time with aspiring or struggling gardeners who need help getting started or rebuilding their confidence.

3. REJUVENATE YOURSELF. Read, browse seed catalogs, laugh, visit, connect, take walks, and do things that help you relax and replenish your energy during these months of cooler days and longer nights.

The first shoots of spring will come soon enough. In fact, many gardeners will be out there planting onion sets in late January and potatoes in February, with a lineup of peas, lettuce, and arugula soon to follow. Put winter to work for you in advance by repairing your garden and, by extension, the world. Without too much work, we can make a difference, while enjoying the rest and rejuvenation we so desperately need.

SEED SAVING

Seed saving is about longevity, heritage, and resilience. It is about creating a legacy, as you pick from the cream of the crop and let these beauties set seed for future generations. You may be thinking that it's hard enough to grow a crop to harvest—why go the extra steps, and use up valuable time and space, to save my own seeds? Here are a few good reasons:

LEARNING: Bolting, or going to seed, is a natural process that plants go through, and by letting it happen, we can deepen our understanding of a crop's life cycle. It is another skill to put in your toolbox, and the rewards are well worth the effort.

FOOD SECURITY: Saving your own seed is yet another step in the direction of becoming totally self-reliant. It also means you don't have to buy seed elsewhere, thereby saving money and ensuring your own food independence.

PRESERVATION: We need to preserve the plant varieties that have been grown for centuries and have been passed down from generation to generation. Saving your own seeds is a way to take direct action against the industrial food system's patenting and genetic modification of seeds. There has never been a more important time in history to be protecting heirloom seeds.

BETTER PERFORMANCE: When you save seeds from the best performers in your garden, you are improving the genetics of the crops that are getting more and more acclimated to your microclimate. Saving seeds strengthens the ability of your crops to fend off disease, pests, and drought, making gardening easier and more abundant year after year.

BETTER TASTE: Most commercial seeds are bred for looks and storage life, so they can survive the long journey in a refrigerated truck. The beauty of a backyard garden is that you can breed for taste and nutrition. Pick the crops that you like best; save their seeds, and each year you will fall more and more in love with your favorite varieties.

IT'S YOUR RIGHT: Many farmers are being sued for saving their own seeds, as large companies try to monopolize the market by forcing farmers to buy their seeds year after year. It is your right to grow and save your own seeds, and I encourage you to exercise that right.

IT'S FUN: Trading seeds with other gardeners is an age-old pastime. It brings people together to share not only the seeds, but also the wealth of stories, growing tips, and recipes that come with them.

TIPS FOR INCREASING YOUR LOCAL FOOD SECURITY

The first time you harvest fresh greens from under a row cover during a snowstorm that shuts down the city, you may have a major "aha" moment when you realize that having your own garden increases your family's food security. When you're finally able to make it to the supermarket and discover the bare shelves, you will see how vulnerable we've become as a society by relying on long-distance growing and transportation of our food. And if you live in a food desert you already know how precious fresh produce is. Here are several proactive steps you can take to increase your local food availability and, thus, food security:

1. SHOP AT YOUR LOCAL FARMERS' MARKET. If you're lucky enough to have a farmers' market nearby, then get out and support your farmers, meet neighbors, and stock the fridge and pantry with healthy, organic, local produce. If you participate in the Supplemental Nutrition Assistance Program (SNAP), you can double your benefits at many farmers' markets through a nonprofit program called Wholesome Wave.

2. SIGN UP FOR A CSA. Joining a CSA (Community Supported Agriculture) farm gets you a prepaid weekly farm box and provides local farmers with up-front "seed money" so that they have a secured customer base for the food that they grow and you eat. Some CSAs offer discounted or even free worker shares for putting in time on the farm or volunteering your home as a delivery location. There are also CSAs that provide donated food shares with the help of private donations and grants.

3. TEACH YOUR CHILDREN TO APPRECIATE HOMEGROWN AND LOCALLY PRODUCED FOOD. In doing so, you will increase their long-term food security by providing them with a skill they will take with them through life. Go to bed satisfied knowing that you are bridging the knowledge gap about growing food to feed our own families as well as generations to come.

4. ADVOCATE FOR CHANGE. Many cities have ordinances that create barriers to increased local food security. No backyard chickens, bees, or goats. No front-yard gardens. No compost piles. No accessory structures, including greenhouses and tool sheds. No commercial sales of produce in a residential neighborhood. The good news? Some cities are now rewriting their ordinances to be more friendly to food-growing activities. Unfortunately this is not yet the norm, and together we need to raise awareness and influence policy to support making our communities more food-secure. You can help do this by inviting local government leaders to your community garden. Educate them about best practices nationwide in urban agriculture. Work with them to frame change not as a "for-or-against" scenario but rather as a way to strengthen our communities.

It seems so simple to plant a seed and grow food. That seed, however, is often a symbol of change. It grows not only food, but power, strength, connections, an outspoken voice, and the confidence that you will always be able to provide for your family.

REST FOR THE SOIL AND THE SOUL

All living things need a rest, and a garden is no different. It is critical that you allow time for both your soil and your soul to rest and rejuvenate. I find the best times to do this are in the heat of summer and the dead of winter. Spring and fall are spent busily turning soil and planting seeds while the weather is ideal. The heat of summer in most places can be a bit oppressive and is a good excuse to do some writing, reading, and napping under the shade of a big oak tree. Things definitely slow down in the winter, making it a great time to relax and reflect on the season past, to flip through seed catalogs and dream about next year's garden.

A great way to let your garden rest and replenish is by planting cover crops and green manures. A cover crop is essentially a crop that is grown to promote soil quality, biodiversity, and wildlife while suppressing weeds, pests and disease. Green manure is created when a cover crop is plowed under to further improve soil fertility, biology, and organic matter. Soil organisms like fungi, worms, and bacteria decompose the vegetative matter and convert it into humus, which improves the quality of the soil. This is extremely valuable for organic gardeners who rely on natural fertilizers to grow healthy crops. Cover crops are nature's savings accounts, earning interest as they grow and rot. They will rebuild organic matter, improve water retention and nutrient-holding capacity, and increase carbon sequestration in the soil. These are significant benefits that can restore productivity and help ensure the long-term health of our planet.

★ TOP COVER CROPS ★

AND GREEN MANURES

Here are some of my favorite cover crops that will make your soil rich and save you money on fertilizer:

COOL SEASON

AUSTRIAN WINTER PEAS: A nitrogen-fixing winter cover crop that can withstand temperatures as low as 10°F, best planted in fall or very early spring. The top few inches of the plant, called pea tips, are a delicacy and can be harvested for salads and stir-fries. I like to plant Austrian winter peas with clover, vetch, and rye in the fall to overwinter, and then plow them in as a green manure in the spring a few weeks after they start flowering.

RED CLOVER: Best planted in fall for spring blooms that can either be harvested for tea or be left to feed the bees and the soil. Best planted with a grass such as oats, rye, or winter wheat and plowed under as a cover crop in the spring after a few weeks of flowering but before it goes to seed.

HAIRY VETCH: A lush nitrogen-fixing plant that can vine up to 12 feet long but rarely gets any taller than 3 feet unless it has another crop to climb. Its beautiful pastel flowers pull nitrogen from the air into the soil and attract bees and other beneficial insects. Its messy mass of vines can be a bit of a hassle to deal with but it is well worth it, as it provides the soil with an abundance of nitrogen and organic matter. Plant in fall with rye, oats, or wheat and plow in as a green manure in spring after a few weeks of flowering.

WINTER RYE: A fast-growing cool-season cover crop that grows well with clover, Austrian winter peas, and hairy vetch. This robust grain has an extensive fibrous root system that breaks up soil compaction and has an overall positive effect on the soil's condition. It can withstand drought, germinate in cold soil temperatures, and tolerate temperatures below 20°F. It works well as a nurse crop for slower-growing legumes like clover and helps provide some support for climbing legumes like hairy vetch. Plant in fall and plow under as a green manure in spring before it goes to seed.

OATS: Not as resistant to cold winters, this small grain makes a good cover crop in rotation with vegetables because it is fast growing and easily killed. Similar to rye, oats are an excellent nurse crop for slower-growing legumes such as clover, vetch, and winter peas. Plant in fall for a winter kill or early spring for green manure.

FORAGE TURNIPS: I like to mix some purple-top turnips in with my cool-season cover crops. It only takes about 10 pounds for an acre, or 3 ounces of seed per 1,000 square feet. That's very little seed but a whole lot of turnips to eat through the winter and early spring. You can also eat the tasty turnip greens all winter so long as they're not covered in snow.

FAVA BEANS (OR BROAD BEANS): These robust nitrogen-fixing, cool-season beans do well when planted in early spring or fall. They are both an excellent cover crop and a gourmet eating bean. Pick when green beans are plump inside the large pods, or mow and plow under after a few weeks of flowering but before bean pods set. I will usually plow in most of the crop and save a small section for eating.

WARM SEASON

BUCKWHEAT: This may be my favorite cover crop, as it provides a quick soil cover that suppresses weeds, improves soil fertility, and attracts bees and beneficial insects. It is especially good for soil deficient in phosphorus, as it is able to scavenge phosphorus along with some calcium from deep within the soil and release these nutrients so they're available for the crops that follow. It also works well as a nurse crop to help establish slow-growing cover crops like clover and alfalfa. It sprouts quickly and creates a humid microclimate, which helps the seeds below to germinate.

SORGHUM-SUDANGRASS: This is an annual grass that is excellent for building up soils, suppressing weeds and nematodes, and breaking up subsoil. It grows from 5 to 12 feet tall and can be mowed at about 3 feet to increase the root mass by forcing roots to grow deeper. It is not frost-tolerant, so plant after last frost. It is widely adapted and will grow just about anywhere in the United States, so long as it gets adequate rain during its establishment phase. It grows very well with iron and clay peas. Mow in fall and plow under as a green manure for a big dose of organic matter.

SOUTHERN PEAS: This heat-loving legume is ideal for hot, humid climates—most of the eastern half of the United States plus the West Coast. It can tolerate drought, heat, and low fertility, making it a low-maintenance option for summer gardens. Plant after last frost through midsummer and plow under as a green manure before planting fall crops. Iron and clay peas are my favorite.

SUNHEMP: Similar to sorghum-Sudangrass, this heat-loving annual provides nitrogen and organic matter to the soil. It can grow 12 feet tall, so I recommend mowing when it reaches 3 to 4 feet, which will increase the bushiness of the plant and its root growth. It grows well with iron and clay peas or all on its lonesome, and it makes a beautiful cut flower. Mow in late summer and plow under for fall planting.

MILLET: This easy-to-grow grain is great as a cover crop mixed with a legume such as sesbania or soybeans. Plant after last frost through midsummer and plow under in fall.

SESBANIA: This annual legume is a big biomass producer, as it can grow up to 12 feet tall in 3 months. It does especially well without much rain, making it a great option for arid climates. Plant in early summer to be plowed under in fall.

★ TOP 5 ★

PERENNIAL HERBS

I love a garden that takes care of itself, and I especially appreciate these five hardy herbs that keep thriving even when I am away for weeks at a time:

1. **ROSEMARY:** A very tasty and medicinal herb that also makes an excellent landscape plant. It is the most vibrant, healthy plant in my front yard; it is completely undemanding and I do nothing but harvest from it. Rosemary does like good drainage, so plant it in an area that stays on the dry side.

2. **OREGANO:** A powerful, antioxidant-rich herb that can liven up many a dish. Of all the herbs I grow, this is the one I pick the most for cooking. Every time I make eggs or cook salmon, for example, a few sprigs of my Italian or Greek oregano (or both) make the cut.

3. **THYME:** A beautiful herb that softens the hard edge of any raised bed or stepping-stone. There are dozens of varieties: some creeping, some upright, and others hanging. The most common culinary variety is English (or winter) thyme. Creeping thyme is great for between cracks in stone walls or around stepping-stones, as it is pleasant to walk on; it is also a great nectar source for bees. Another variety that I like to cook with and make tea from is lemon thyme, which is a wonderful addition to many dishes. Like most herbs, thyme prefers dry feet, so plant it in an area with good drainage.

4. **PARSLEY:** An amazing herb that has lots of health benefits, with its high levels of antioxidants, vitamins, and folic acid. It is also a wonderful habitat plant for swallowtail butterflies, bees, and birds. It grows best in moist, well-drained soil, and while a little slow to germinate from seed, it grows rapidly once established. It is the most productive herb in my garden throughout the winter, bar none. I like to juice it and chop it up in salads.

5. **CHIVES:** Both onion and garlic chives are wonderful plants to have in the perennial herb garden and around the landscape. They like fertile soil and appreciate getting frequent haircuts. While grown primarily for their leaves, they both have potent edible flowers that make for a beautiful and tasty addition to many dishes. They are also both beneficial in the garden as they help to repel bad bugs and provide a habitat for bees.

★ TOP 5 ★

PERENNIAL VEGETABLES

For those of you who are rooted firmly where you are, these perennials will produce year after year without your having to replant.

1. **ASPARAGUS:** Prepare a bed approximately 4 feet wide by removing all weeds and incorporating a heap of compost into the top 6 inches of soil. Plant one-year-old crowns in fertile, well-drained soil in full sun and pick a spot where you won't mind having asparagus for the next twenty years. Dig trenches about 12 inches wide and 6 inches deep, and plant crowns into the trenches 18 to 24 inches apart, ideally after soaking them in compost tea for 20 minutes. Cover with a few inches of soil and continue to add more soil as the bed settles to keep it a little higher than the surface for good drainage.

2. **RHUBARB:** Grow in full sun unless you are in a very hot, dry climate—then plant where it can get some afternoon shade. Rhubarb does best in a cool climate that has a substantial winter. Add plenty of compost before planting crowns in early spring or fall and fertilize every year by adding a few inches of compost as a side dressing in early spring.

3. **HORSERADISH:** A cold-hardy plant that prefers full sun but can also tolerate partial shade. Plant from roots in well-drained, fertile soil (with a pH between 6.2 and 6.8) into 6-inch-deep furrows amended with compost. Plant root cuttings about 1 foot apart at a 45-degree angle with the top of the cutting 2 inches below the soil line. It is a very rugged plant that will need minimal maintenance, though it will do best with a weekly watering and a few inches of mulch around its base.

4. **YACÓN:** A vegetable from the Andes that is easy to grow and has many health benefits. Similar to the Jerusalem artichoke, another excellent edible perennial, yacón is a rhizome with edible succulent roots attached. It grows into a 6-foot-tall shrub and is easily propagated by small tubers that are plucked off the main tubers at harvest.

5. **TREE COLLARDS:** Basically these are perennial collards that do well in cooler climates. They can be left to spread out over the ground or trellised to save space. They can be propagated by cutting, which is best done before they get too woody, at about 3 years of age. Prune heavily to get a bushier plant with more leaves to harvest. While tree collards can grow for ten years or longer, I recommend growing them for 3 to 5 years and replanting, as they tend to get woody and unwieldy with age.

VIRTUES IN ACTION

1. By practicing good stewardship through composting, your soil will become more fertile and your soul will feel more fulfilled.

2. By having a well-conceived plan and a clear vision, your plants will feel less stressed and your life will feel more productive and purposeful.

3. By taking the initiative to till and tend the soil, you will become more confident and empowered to take action in other areas of your life.

4. By sowing seeds, you will learn to have faith that even the tiniest action on your part can make a big difference.

5. By having patience through the sometimes painful stages of growth, you will learn that the best things in life are always worth the long wait.

6. By healing sick plants and animals, you will cultivate compassion for others.

7. By reaping the rewards of the harvest, you will learn to practice gratitude in all aspects of life.

8. By sharing your garden with others, you will develop a more generous spirit.

9. Finally, and most importantly, by persevering through the challenges that growing healthy food presents, you will be able not only to provide better for your family, but also to help foster a more sustainable world.

REVIVING A LEGACY

If you're into organic gardening, or interested in learning more about it, thank you. You may not realize how much we need your body of knowledge. Until World War II, just about everyone knew how to grow food in one way or another. But the war changed American culture. Wartime chemicals and engineering innovations birthed the industrial farming movement; booming business provided jobs for returning servicemen in cities and factories; more women entered the workforce; and convenience foods started taking center stage on dinner tables.

In short, over the past sixty years or so, traditional food-growing wisdom has skipped two generations, if not more. Those who have begun gardening recently are helping to refill this void. Today's citizen farmers are reviving an age-old legacy, helping to restore critical life-sustaining, future-ensuring knowledge to humanity.

Thank you for joining us.

FARMER D'S TOOL CHEST

The following list is a collection of useful websites to help you grow. While most of these are national sites, be sure to also seek out local resources such as organic and urban farming organizations. Thankfully, there are so many amazing organizations and individuals contributing to the movement that I have barely scratched the surface with this list.

GARDENING SUPPLIES

I recommend first checking your local organic garden center, feed-and-seed store, and farmers' market for your gardening needs. If you don't find what you are looking for locally, I assure you the following websites will provide you with access to everything you could ever need to support a healthy garden and sustainable lifestyle.

Clean Air Gardening (Texas): www.cleanairgardening.com
Eartheasy (British Columbia): www.eartheasy.com
Farmer D Organics (Atlanta): www.farmerd.com
Garden Tool Company (Colorado):
    www.gardentoolcompany.com
Gardener's Supply Company (Vermont): www.gardeners.com
Gardens Alive! (Indiana): www.gardensalive.com
Johnny's Selected Seeds (Vermont): www.johnnyseeds.com
Planet Natural Garden Supply (Montana):
    www.planetnatural.com
Territorial Seed Company (Oregon): www.territorialseed.com
Williams-Sonoma Agrarian (California):
    www.williams-sonoma.com/agrarian

FARM SUPPLIES

While these online stores are geared for the small farmer and market gardener, they offer great options for the home grower as well.

FarmTek (Iowa): www.farmtek.com
Harmony Farm Supply (California): www.harmonyfarm.com
Peaceful Valley Farm Supply & Nursery (California):
    www.groworganic.com
Seven Springs Farm (Virginia): www.7springsfarm.com

BENEFICIAL ORGANISMS FOR PEST CONTROL

Check your local garden centers and hyrdroponic stores for beneficial insects, or order online. Here are a few options.

Arbico Organics (Arizona): www.arbico-organics.com
Buglogical Control Systems (Arizona): www.buglogical.com
Gardens Alive! (Indiana): www.gardensalive.com
Planet Natural (Montana): www.planetnatural.com
Tip Top Bio-Control—wholesale only (California):
    www.tiptopbio.com

SEEDS AND PLANTS

There are far too many great seed companies to list them all, but here are some of my favorites.

Baker Creek Heirloom Seeds (Missouri and California):
    www.rareseeds.com
Botanical Interests (Colorado): www.botanicalinterests.com
Burpee Seeds and Plants: www.burpee.com
The Cook's Garden (Pennsylvania): www.cooksgarden.com
The Cottage Gardener (Ontario, Canada):
    www.cottagegardener.com
Fedco Seeds (Maine): www.fedcoseeds.com
Filaree Garlic Farm (Washington): www.filareefarm.com
The Garlic Store (Colorado): www.thegarlicstore.com
High Mowing Organic Seeds (Vermont):
    www.highmowingseeds.com
Hudson Valley Seed Library (New York):
    www.seedlibrary.org
Johnny's Selected Seeds (Vermont): www.johnnyseeds.com
Living Seed Company (California):
    www.livingseedcompany.com

Potato Garden (Colorado): www.potatogarden.com

Seed Savers Exchange (Iowa): www.seedsavers.org

Seeds of Change (California): www.seedsofchange.com

Southern Exposure Seed Exchange (Virginia):
    www.southernexposure.com

Sow True Seed (North Carolina): www.sowtrueseed.com

Territorial Seed Company (Oregon): www.territorialseed.com

Totally Tomatoes (Wisconsin): www.totallytomato.com

Turtle Tree Seeds—biodynamic and nonprofit (New York):
    www.turtletreeseed.org

ORGANIC FEED SUPPLIERS

Given the heavy weight of feeds, I recommend sourcing locally, either directly from the mill, pet or feed-and-seed store, or garden center. Here are some additional sources.

Countryside Organics (Virginia):
    www.countrysideorganics.com

Coyote Creek Farm (Texas): www.coyotecreekfarm.com

Green Mountain Feeds (Vermont):
    www.greenmountainfeeds.com

Hiland Naturals (Ohio): www.hilandnaturals.com

Reedy Fork Dairy Farm (North Carolina):
    www.reedyforkfarm.com

EQUIPMENT

Here are a few resources for walk-behind tractors, implements, and other equipment:

Earth Tools (Kentucky): www.earthtoolsbcs.com

Ferrari Tractors (California): www.ferrari-tractors.com

Market Farm Implement (Pennsylvania):
    www.marketfarm.com

Peaceful Valley Farm & Garden Supply (California):
    www.groworganic.com

PUBLICATIONS AND ONLINE RESOURCES

American Community Gardening Association:
    www.communitygarden.org

ATTRA (National Sustainable Agriculture Information
    Service): www.attra.org

Beginning Farmers: www.beginningfarmers.org

Biodynamics Journal: www.biodynamics.com/journal

Compost Resource: www.howtocompost.org

Crop Mob: www.cropmob.org

Eat Well Guide: www.eatwellguide.org

Food Routes: www.foodroutes.org

Growing for Market: www.growingformarket.com

Heifer International: www.heifer.org

Heritage Radio Network: www.heritageradionetwork.com

Local Harvest: www.localharvest.org

Modern Farmer Magazine: www.modfarm.tumblr.com

Mother Earth News: www.motherearthnews.com

Organic Gardening Magazine: www.organicgardening.com

Organic Seed Alliance: www.seedalliance.com

Permaculture Magazine: www.permaculture.co.uk

Quantum Agriculture: www.quantumagriculture.com

Rodale Institute: www.rodaleinstitute.org

Slow Food USA: www.slowfoodusa.org

Sustainable Table: www.sustainabletable.org

Urban Farm Magazine: www.urbanfarmingonline.com

FORUMS

Biodynamics: www.biodynamics.com/forum

Dave's Garden: www.davesgarden.com

Garden Stew: www.gardenstew.com

Garden Web: www.gardenweb.com

The Helpful Gardener: www.helpfulgardener.com

Kitchen Gardeners International: www.kgi.org

Organic Gardening: forums.organicgardening.com

Permaculture Forums: www.permies.com/forum

BIODYNAMIC RESOURCES

Biodynamic Farming & Gardening Association:
    www.biodynamics.com

Cityfood Growers: cityfoodgrowers.com.au

Demeter Association, Inc.: www.demeter-usa.org

Michael Fields Agricultural Institute (Wisconsin):
    www.michaelfields.org

Rudolf Steiner College (California):
    www.steinercollege.edu/biodynamics

ACKNOWLEDGMENTS

I think the only person who wanted this book done more than me was my patient, loving, and supportive wife, Stephanie. I am forever grateful to her for putting up with all the late nights and for keeping our son, Tilden, from attacking the computer as I wrote. You are the best thing that has ever happened to me and I love you with all my heart. May your hard work with To-Go Ware and all the other ventures in life make you happy and successful beyond your wildest dreams.

The people who are most responsible for helping me to achieve my goals both personally and professionally are my loving parents, Avril and Stanley Joffe. Thank you both for teaching me how to be a mensch and encouraging me to follow my dreams. I am forever grateful that you have never stopped believing in me and have supported all of my crazy adventures. You are the best role models a new parent could ever wish for. Not only did Dad support me over the past 36 years of my life, helping nurture his homegrown farmer, but he has also played a huge role in helping run our business. Thanks, Dad, for holding it all down while I spent so many hours in the book bubble. I am beyond blessed to have such a wise, grounded, and loving father and business partner.

We all owe a huge thank-you to Susan Puckett for helping make this book what it is. There is no way I would have been able to finish this monumental task without her incredible writing skills and encouraging spirit. Susan, I cannot thank you enough for all the time and passion you have so lovingly dedicated to bringing this from seed to harvest. I am sincerely grateful for you and am looking forward to many more adventures together.

A special thanks goes out to Rinne Allen for her incredible passion and dedication to capturing the essence of the book through her amazing photography. It was such a pleasure working with you and getting to know your warm and generous spirit. Thank you for your creative inspiration and support!

A very special thanks to Pattie Baker, whose years of writing for my CSA newsletters and blog provided a great deal of content that I have been able to share with you in this book. Thank you, Pattie, for your inspiring fountain of energy

and dedication to the movement and for the many hours you have invested in helping me spread the word. A special thanks to you for inspiring the Parable of the Citizen Farmer!

A big thanks goes out to Deborah Geering for making valuable suggestions and testing recipes.

A very special thanks goes out to Amy Hughes for being so supportive in helping cultivate the vision for this book and ultimately for getting it into the right hands with Stewart, Tabori & Chang (STC). I can't say enough how grateful I am for all the incredible support that Dervla Kelly and Emily Albarillo of STC provided in the way of visioning, support, and editing. A special thanks to Dervla for encouraging me to focus on the community side of urban agriculture, which really set the tone and inspired the ethos of the book.

A hearty thanks to DMA and Aaron Spiewak for encouraging me to write the book and for introducing me to all the right people, including David McCormick. A big thanks to David for steering me away from a gardening cookbook and seeing the bigger opportunity to tell my story and share my passion.

This book is in honor of my loving grandparents, Papa Benny, Granny Tilly, Grandma Cissy, and Papa Solly, who all inspired me in their own way. It was a blessing to have all four of my grandparents healthy and strong well into my thirties. I owe much of my passion for food, family, and community to them and this book is in their memory.

This book is also in honor of my son, Tilden Gray Joffe, who is the greatest seed I have ever planted. Nothing has brought me as much joy in such a short time as he has. I look forward to the day he can read it and learn about his papa's roots. It is also in honor of my unborn daughter, who will be 2 months old when the book is first printed.

I am extremely grateful for my inspiring sister, Cindy, who continues to amaze me with her culinary and parenting skills. I am also thankful to her husband, Jason, and their three children, Dylan, Jonah, and Chloe, for all they do to make our family gatherings full of laughter and love. A hearty thanks goes out to my in-laws, Carolyn and Howie

Bernstein, for raising the love of my life and giving us the wedding of our dreams. A big thanks to my wonderful sister and brother-in-law, Royce and Chris Murray, and their beautiful girls, Siena and Gabriela.

I also want to extend a very special thanks to all of the Farmer D staff for holding everything together during the two years I spent working on this book. I cannot express how grateful I am for all of you who put out fires, blazed forward, and kept the ship at sail while I was busy writing. An especially big shout out to rock stars Katie Pigott and Joshua Tabor!

I owe a huge thank-you to Greg Ramsey of Village Habitat Design and Sean Murphy of B+C Studio for being amazing design partners and dear friends. Most of the designs showcased throughout the book were created in collaboration with them. I continue to learn so much from both of you and greatly value your friendship and the amazing support you've shown for Farmer D Consulting over the years. I look forward to designing many more meaningful farms and gardens with you both!

The Farmer D brand owes a big thank-you to Robert Rausch of Gas Studios for all his help over the years with the logo and branding. A warm thank-you also goes out to Lindsey "Cubby" West for her creative input over the years.

Thanks to all the friends along the way who have helped guide and teach me. I am so much a reflection of you all and am grateful for your ongoing support and inspiration. A shout out to James Welch, Ronny Bell, Jason Mann, David Amato, Marcus Rubenstein, Jo and Alain Eagles, Eric Fenster and Ari Derfel, Max Rohn, Carrie Branovan and family, Udi Lazimy, Adam Berman, Stephen Brooks, Mitch Braff, Jeff Vantosh, Jon Rosenthal, Gauruv and Naruna, Grady Cousins, Kyla Zaro-Moore, Steven Cayre, Mohammed Nuru, Jim Fullmer, Robert Karp, Jennifer Levison, Marla Hoppenfeld, Laura Turner and Rutherford Seydel, and the many others who have helped me grow over the years.

A special thanks goes out to Mike and Karen Smith of Longwood Plantation for being my compost partners for all these years and for opening their hearts and minds to me and my biodynamic ideas.

A very fertile thank-you goes out to the amazing people at Whole Foods who helped make the compost dream a reality, especially John Walker, Sandy Pilgrim, Mike Hardy, and Cheryl, Emily, and Darrah of the marketing team.

A big thank-you to the Joshua Venture Group for providing me with an amazing two years of social entrepreneurship training and tremendous support on my journey.

A hearty thanks to the whole team at Williams-Sonoma Agrarian, especially Allison O'Connor, Leah Ducker, and Carmine Fiore, for a wonderful partnership.

Thank you to all my clients who had to wait longer than usual to get responses, reports, and designs during the months leading up to the book deadline. Thank you for understanding and for the opportunity to grow with you! A big thanks to Steve and Marie Nygren of Serenbe; Bob and Kim of Natirar; Bill and Renee Cole; the kind people over at Valley Crest, especially Skip Hogan and Kurt Buxton; Keith and Jay Howard; Doug Goff at the Johnson Development Corporation; Tom Bacus and Josiah Cain at Sherwood; Dave Howerton, Tim Slattery, and Jim Tinson at Hart Howerton; Lloyd Blue and Rick Borst at Longleaf Preserve; Ed Mitchell at Honeywood Farm; Rawson Haverty Jr. and Rodney Cook at historic Mims Park; Lee Walton and Ron Huffman of AMEC; Bill and Dawn Davidson; Jim Farley and the incredible team at Leichtag; Chip Conley; Sarah Brightwood of Rancho La Puerta.

A very heart-felt thanks to Camp Twin Lakes for the amazing work they do every day and for allowing me the opportunity to grow with them and take photos for the book of campers in action on the farm.

A big thanks to all the wonderful people I have been fortunate to work with at Rancho Mission Viejo, Palmetto Bluff, Hampton Island, Playa Todos, Talisker, Sea Island, Newhall, Camelot Homes, The Erdman Group, Daybreak, Mandarin Oriental Hotel, The W Atlanta, Captain Planet Foundation, Children's Healthcare of Atlanta, Robson Communities, the city of Suwanee; my fellow board members at Georgia Organics; and my Biodynamic Association peers.

I owe a huge thank-you to my first farming mentor, Greg David, for setting me on my course. Greg taught me many things about farming for which I am grateful, but more than anything he inspired me to cultivate a vision and pursue it with faith.

I am very grateful for the many valuable lessons I learned from Hugh Lovel, both on his biodynamic farm and in the kitchen.

To all the farmers, chefs, gardeners, conscious consumers, authors, and artists who are making a difference in their community, one plant, meal, and moment at a time.

INDEX

A

air element, 14, 99
animals
    as healers, 130
    planning for, 66–68
annuals
    for attracting beneficial insects, *138*,
        140–41
    for edible landscaping, 71
ants, 44
arbors, 82, *83*

B

Bacillus thuringiensis (Bt), 135
backyard garden, tools for, *28*, *29*
BD prep. *See* biodynamic preparations
beneficial organisms, 135–36
biodynamic agriculture, 13–15
biodynamic preparations (BD prep), 14
    compost pile addition of, 42–43
bokashi, 34
boosters, for composting, 41
brassicas, 111–12, *111–12*
Bt. *See Bacillus thuringiensis*

C

calcium, earthly polarity and, 14
carbon, in compost, 36, 38, 42
citizen farmer basics, 17–19
community composting, 48
community garden
    animals for, 68
    getting involved in, 66, *67*
    planning of, 63–65
    tools for, *28*, *29*

Community Supported Agriculture
    (CSA)
    garden planning for, 63
    starting, 201
companion planting, 114–15, *114–15*
compost
    addition to beds of, 77
    applications of, 20
    buying, 44, *45*, 46
    completed, 44
    ingredients for, 38–39
    making, way of life, 17
    plants for making, 49
    recipe for, 36
    use of, 46
composting
    boosters for, 41
    business opportunity with, 199
    community, 48
    locations for, 32, *33*
    with pigs, 24
    process of, 37
    systems for, 32–36
    tools for, 30–31
compost meditation, 26, *27*
compost pile
    adding biodynamic preparations to,
        42–43
    building of, 41–42
    tending to, 43
    troubleshooting, 43–44
compost tea, 46–47
constellations, four elements and, 99
container gardens, 81
cool-season crops, 112
cosmic force, 14
cover crops, 46–48, 214–15

critter-proofing, 117–18, *119*
critters, with composting, 32
crop rotation, 110–12, *111–12*
crops
    for cutting grocery bills, 162–63
    easiest to grow, 124–25
    by family/group, 112, *112*
    for kids, 194
    by nutrient needs, 113, *113*
    to preserve, 206
    by season, 112
    for sharing, 193
CSA. *See* Community Supported
    Agriculture
culinary herbs, 70, 158, *159*, 215
curbside composting, 35

D

daily observation, *108*, 109
damping off, 94–95
deer, proofing against, 118
diseases. *See* pests and diseases
double-dig, 81
drainage, for composting, 32

E

earth element, 14, 99
earthly polarity, 14
edible landscaping, 70–71, 86
electric composters, 34
entre-manurial spirit, 199–202
exercise while gardening, *84*, 85

F

fall garden, 207
family/group of crops, 112, *112*
Farm Nibbles, 164–65, *165–66*, 167

fire element, 14, 99
foliar feeding, 117
fruit trees, 70, 161

G
gardener's medicine chest, 144–48, *146*
gardens
    for different venues, 60, *61*, 62–63
    fall, 207
    for healing, 130, *131*
    putting fun into, *122*, 123
    sowing directly in, 95–96
garlic, 136, 148, *149*
green manuring, 46–48, 214–15
growing, 104
    business opportunity with, 200–201
    companion planting, 114–15, *114–15*
    critter-proofing, 117–18, *119*
    crop rotation, 110–11, *111*
    extending season for, 208, *209*
    foliar feeding, 117
    tips for, *108*, 109–10
    tools for, 106–7
    weeds, 118–20
grub composting, 34–35

H
harvest, 150
    business opportunity with, 201
    culinary herbs, 158, *159*
    Farm Nibbles, 164–65, *165–66*, 167
    fast food from garden, 160
    impromptu party, 168, *169*
    meditation on, *152*, 153
    tips for, *156*, 157
    tools for, 154
healing, 127
    business opportunity with, 201
    gardener's medicine chest, 144–48, *146*
    gardens for, 130, *131*
    with garlic, 148, *149*

grow juice, 142–43
    herbal meditation, *128*, 129
    plants for attracting beneficial insects, *138*, 139–41
    year-round, 141–42
heavy feeders, 113, *113*
heavy givers, 113, *113*
heavy-metals test, 76
herbal meditation, *128*, 129
home garden
    animals for, 68
    planning for, 60
horn manure (BD prep 500), 14
horn silica (BD prep 501), 14
hugelkultur, 35, 81
humanure, 35–36

I
indoor garden, tools for, *28*, 29

L
large-scale composting, 35
legumes, 111–12, *111–12*
light feeders, 113, *113*
lime, earthly polarity and, 14
living walls, 82
loam, 77
local food security, 212–13
location, selection of, 75

M
maggots, 44
manure, applications for, *40*, 41
market farm
    garden planning for, 63
    tools for, *28*, 29
massaged kale slaw, *170*, 171
medium feeders, 113, *113*
microclimates, 75
model citizen plants, 115, *115*
moon, planting with, 96, *97*, 99
mulching, 109

N
neutral ingredients, sources of, 38
nightshades, 111–12, *111–12*
nitrogen, in compost, 36, 38, 42
no-till, 81
nutrients
    crops by, 113, *113*
    management of, 110
nut trees, 70, 161

O
odor, with composting, 32, 44
onions, 111–12, *111–12*
open-pile system, 32
oxygen, compost function of, 36

P
perennials
    for attracting beneficial insects, *138*, 139–40
    herbs, 215
    to replace ornamentals, 70
    vegetables, 216
pests and diseases, 132–35
    control of, 110, 135–39
picking, 109
planning, 51
    for animals, 66–68
    business opportunity with, 199–200
    community garden, 63–65
    defining intentions, 57
    first farm, 51–52
    gardens for different venues, 60, *61*, 62–63
    steps for, 55–56, 75–77
    tools for, 58–59
    visioning, 56–57
planning meditation, 53
planting beds, 76–77

plants
    for attracting beneficial insects, *138*,
        139–41
    for compost production, 49
preserving
    crops for, 206
    tools for, *204*, 205
pruning, 110

R
rabbits, 41, 117
reaping. *See* harvest
rodents, dealing with, 44
root crops, 87, 111–12, *111–12*

S
seasonal tasks, 112, *112*
seedling trays, 90, *92*, 93–94
seeds, 93
    easiest to grow directly in ground,
        102, *103*
seed saving, *210*, 211–12
shade, for composting, 32
sharing, 175–76
    building eleven gardens in one day,
        178, 179
    business opportunity with, 201–2
    crops for, 193
    crops for kids, 194
    from farm to city, 176
    growing for those in need, 184–85
    meditations on, 182
    order of seeds, 184
    plant gifts, 185
    school garden challenges, 188–89
    take garden viral, 187–88
    tools for, 180
    working with children, 190, *191*, 192
shucking meditation, 203
side dressing, with compost, 46
silica, cosmic force and, 14
soaking seeds, 93

soil
    addition to beds of, 77
    builders of, 113, *113*
    compost and, 20
    compost pile building with, 42
    for seedling tray, *92*, 93–94
    testing of, 75–76
soil erosion, reducing, 110
sowing, 89
    business opportunity with, 200
    directly in garden, 95–96
    easiest to grow directly in ground,
        102, *103*
    with moon, 96, *97*, 99
    for quicker returns, 99
    seeds of change in community, 100,
        101
    tools for, 90–91
    in trays, *92*, 93–94
summer vegetable gratin, 181
sun, in biodynamic agriculture, 14
sustaining, 196, 199
    business opportunity with, 202
    extending growing season, 208, *209*
    increasing local food security,
        212–13
    rest for soil, 213
    reviving legacy, 217
    seed saving, *210*, 211–12
    tools, 202–3
    virtues in action, 217

T
temperature, achieving, 43
three-bin system, 34
tilling
    alternatives to, 81–82
    business opportunity with, 200
    tools for, 78–79
tools
    commonly used, *28*, 29
    for composting, 30–31

    for growing, 106–7
    for harvest, 154
    for planning, 58–59
    for preserving, 205
    for sharing, 180
    for sowing, 90–91
    sustaining, 202–3
    for tilling, 78–79
transplanting, 96
trellises, 82, *83*
trellising, *108*, 109
tubers, 87
tumbler, 34

V
vermicompost, 34
vertical gardening, 82, *83*
vinegar, 120
visioning, 56–57

W
warm-season crops, 112
water, for composting, 32, 36
water element, 14, 99
watering, seed trays, 94
weeds, 118–20
whippoorwill pea and butternut squash
        stew, 172, *173*
window gardens, 82
wire bin, for composting, 32–34
wood, for planting beds, 76–77
wood bin, for composting, 32–34
worm bins, 34, 44
worm farming, 199